Praise for *The Ten Commandments for Effective Standards*

"Karen is the best in her field. Candid and concise, 'The Ten Commandments for Effective Standards' is a must read for individuals interested in developing industry standards. Packed with wisdom and practical insights, readers will be able to apply the learnings!"
Shrenik Mehta, Chairman, Accellera; and Former Senior Director, Microelectronics, Sun Microsystems

"Karen does an excellent job in illuminating the sometimes murky workings of standards groups. In espousing her 'commandments,' she identifies the often innocuous situations that have caused major issues in the standards world and how to avoid those types of issues in the first place. This book should be required reading for anyone who will be participating in a standards organization, both inside and outside the EDA world."
Stan Krolikoski, Group Director, Cadence Design Systems; and Chair, IEEE Design Automation Standards Committee

"So, what's the big deal about 'The Ten Commandments for Effective Standards?' First, the author, Karen Bartleson, is a well-respected and recognized standards guru, particularly in technology circles. She understands the ins and outs of standards history, best practices, business implications, and current issues. She really knows what she's talking about. Second, Karen is incredibly gifted in conveying when and why standards are necessary and how to get the most out of them. She writes in a clear and entertaining style, plus she's passionate. Finally, Karen's able to ground the discussion with real-world examples and fun cartoons. This is a must read for anyone participating in standards development!"
David Lin, Vice President of Marketing, Denali Software

G000160625

"Karen Bartleson makes it clear that there are no gods or goddesses in standards, but she neglects to mention that she is the Queen of EDA Standards. It's a dirty job, but Karen makes life much easier for any standards committee she belongs to. And now she has written a hands-on and easy-to-read primer for those engineers who wish to enter the important area of standards development."
Gary Smith, Chief Analyst, Gary Smith EDA

"Standards open the road for a community to remove barriers and progress more rapidly. They create the shared highways for all players—partners and competitors alike—to travel faster. Building on the greater combined wisdom and innovation of the community, standards enable whole industries to evolve quickly, benefiting not only their participating companies, but customers as well.

Karen provides a vivid insider view of the world of EDA standards creation, capturing this complex process with clear narrative, often punctuated with humor. She supports her writing with real industry examples. While many potential views and perspectives may be argued on particular cases, we should welcome the discourse on standards and their effective construction."
Gadi Singer, Vice President & General Manager, System-on-Chip Enabling Group, Intel Corporation

"The EDA Technical Committee (EDA-TC) was formed in JEITA in 1990, and aims to drive initiatives of international EDA standardization in Japan. Since then, the EDA-TC has been actively contributing to design language standardization, which most electronic design engineers are currently using worldwide. This year, we celebrate the twentieth anniversary of its birth.

Karen is one of some active veterans in the EDA standards community, contributing to EDA standards and also being very supportive to EDA-TC activities. I believe this book makes relevant people understand the real value of standards. She will lead them to share her experiences and perspectives through her Ten Commandments. This book is what I really want to have, and I highly recommend it."
Satoshi Kojima, Fellow of EDA Technical Committee, JEITA

"If you, as a business or individual, are trying to understand whether to participate in standards or understand their value, you will undoubtedly be struggling with many questions, questions that are not readily answered by reading literature from standards development organizations. You will be hungry for experience from a veteran's perspective that is clear to the novice. Karen is the person with that experience. The personal experience and insight she has woven into this book will help you kick start your adventure in standardization. I only wish I had had this kind of insight and guidance when I started out. 'The Ten Commandments for Effective Standards' is simply missing one important commandment: 'Thou shalt read this book before engaging in standards.'

I highly recommend reading this book rather than learning the hard way. Standardization has many challenges and issues; you must walk into that process with your eyes wide open. 'The Ten Commandments for Effective Standards' is an essential read."
Jeff Green, Senior Vice President, McAfee Labs, McAfee

"'The Ten Commandments for Effective Standards' is a great companion to the standards developer and for standards committee leadership. It is an even better companion to those who wait for and use standards, as it serves to remind adopters that there are many facets to standards development that need to be addressed to get to a completed standard. This helps explain why it often takes so much time for some standards and why it is fast for others. If you follow the commandments you may have a smoother and quicker path to standards success. I expect standards developer conversations to begin to use more of the terms 'The Ten Commandments for Effective Standards' defines to usher further rapid development of business-relevant standards. As all good standards evolve, I look to Karen to be on the lookout for additional commandments that might be worthy to join the ten she has detailed here."
Dennis B. Brophy, Director of Strategic Business Development, Mentor Graphics Corporation

"Karen Bartleson is a brilliant and friendly IEEE-SA CAG member. This book introduces practical and useful information for readers with suitable past case studies. As Karen mentions in this book, standardization is sometimes called a 'war.' Readers will understand the severe aspects of standardization processes in reading the fourth and fifth 'commandments.' However, readers will find Karen's friendly smiling face between the lines and will notice other positive aspects of standardization to stimulate the development of new technologies and contribute to market expansion. After reading this small book, I believe that the readers will have the motivation to attend and contribute to standards activities."

Shinji Tanabe, Ph.D., Chief Researcher of Mitsubishi Electric Corporation, IEEE Senior Member, IEEE-SA CAG (Corporate Advisory Group) Member

"I remember the day I first met Karen Bartleson. She had called my office inquiring about being a Synopsys blogger. Usually, it takes time for people to grasp the difference between inbound and traditional marketing, but Karen got it right away. Not only had she kept up with my explanation, I distinctly remember her questions remaining one step ahead of my answers. Within the week, Karen had launched her blog, The Standards Game.

I've always wondered how Karen grasped the principles of new media so quickly that day-a mystery that was solved after I read 'The Ten Commandments for Effective Standards.' Creating an effective standard requires the same fundamental concepts as creating effective new media. Both involve communication. Both involve people and the myriad of complications that come with them. And, lastly, both require a laser-like focus on the long-term objective: to serve the end-customer before either the standard itself or the corporate politics.

If you're interested in establishing standards within your industry, this little book is the place to start."

Ron Ploof, Executive New Media Strategist, OC New Media; and Author of *Read This First: The Executive's Guide to New Media-From Blogs to Social Networks*

"STARC actively promotes standards in the EDA industry, and this book is an excellent roadmap for anyone wishing to participate in the standards process. Especially in the fast-paced semiconductor industry, following the advice in this book can only accelerate the development of timely, relevant standards which are essential to a healthy EDA ecosystem."

Nobuyuki Nishiguchi, Vice President and General Manager, Semiconductor Technology Academic Research Center (STARC)

"If you develop technical standards, or if you want to know more about how technical standards are made and what goes into making and maintaining an international technology-based standard, you should read this book.

Karen Bartleson has managed to put over fifteen years of intensive standards work experience into ten simple-to-understand rules and recommendations. You will see why it is important to be as open as possible, but also that mixing patents and standards requires caution, and how openness among participants is critical to standards' success, but also that, at the same time, everyone has an agenda, and (almost) everyone can be your competition.

This is an optimistic book! It takes the view that standards are good and that developing a technology standard is a worthwhile investment in time and money. I have worked with Karen on the development of several standards, and I am sure that you will find her book useful and entertaining."

Oz Levia, Vice President of Marketing, Springsoft

"STMicroelectronics is a leading supplier of electronic chips. Our success as a company relies on the existence of many standards for our products to be successful on the market, standards such as GSM, WIFI, Bluetooth, MPEG, H264, Java. For the successful and efficient design of our products, we also rely on the existence and effective adoption of many EDA standards, such as the ones promoted by Karen in her 'real life,' standards such as Verilog, VHDL, LEF, DEF, SystemC, TLM, IP-XACT, UPF, and CPF. There are many origins for standards: de facto standards, ad hoc standards, standards evolved from open-university work, or former proprietary. Whatever the genesis of a given standard, its success and its value depend on the few parameters that Karen describes in a lively manner in her book: openness, technical relevance, pragmatism, appropriate timing, etc.

I sincerely hope that this book will become bedside reading for many participants in standards bodies, in the electronics and EDA industries. It will enable us to all speak the same language and to understand the rationale for participation of all participants to standards. Of course, a participant to a standards body has both a technical motive as well as a business motive. Some of these motives may be visible and others not. Through Karen's wide experience and meaningful anecdotes, this book enables any participant to standards to understand the mechanics and politics of what works and what doesn't and why. I hope to see many standards happen faster in the future, thanks to this guide."
Philippe Magarshack, Group Vice President, Technology R&D, Design Solutions GM, STMicroelectronics

"I have been working with Karen on standards for over two decades. Our extreme success in the standards area is a result of applying Karen's 'Ten Commandments for Effective Standards.' Anyone wanting to find success in the challenging area of standards setting should start by reading this book. It is an indispensible guide to the dos and don'ts of standards setting. Read the book and get started on success. Or ignore the book, and regret it later."
Rich Goldman, Vice President, Corporate Marketing and Strategic Alliances, Synopsys

"Karen provides a much-needed analysis of and invaluable lessons learned in the process of birthing technical standards. This work provides a clear guide for overcoming the serious pitfalls that have plagued standardization efforts in the past. Karen's practical Ten Commandments are based on time-tested real-world experience and should be mandatory reading for anyone collaborating with multiple suppliers seeking to bring them together to cooperate on any kind of a standard. Karen has eloquently blended together historical case studies with recent technical standardization experiences that not only generally educate the reader but also provide very specific practical guidance regarding avoiding the legal and ethical challenges that can seriously harm corporate intellectual property positions and/or significantly slow valuable industry standardization efforts. No question that future standards, especially future EDA standards, will be more likely to become reality after careful study of this book. Well done, Karen."
David Peterman, Wireless EDA, Texas Instruments

"It is a safe bet that each of you picking up this book at some point in your life must have been confronted by what you consider a 'good' or a 'bad' standard. We all agree that standards are an essential component of making diverse things (or people) interact harmoniously and, if done well, can accelerate development in a certain field and lead to better outcomes for all. In contrast, ill-conceived standards most often lead to stagnation, fragmentation, and ultimately inferior results.

I truly wish that anyone involved in the creation of standards in any field reads Karen's book very carefully. Based on her experience in the field of electronic design automation, Karen has synthesized a clear and transparent set of guidelines on what it takes to create 'good standards.' I couldn't agree more with her insights. My only wish is that the book should have been available many years earlier. It would have avoided a lot of wasted time and misguided efforts."
Jan Rabaey, Donald O. Pederson Distinguished Professor, Electrical Engineering and Computer Science, University of California at Berkeley

A Note from the Publisher

We're excited to introduce *The Ten Commandments for Effective Standards* as the first book in the Synopsys Press Business Series. Author Karen Bartleson brings to this practical guide not only extensive professional experience delivered in an easy-to-read style, but also the experience of many other industry experts through significant peer review and input. You can follow Karen's blog, The Standards Game, at http://www.synopsys.com/blogs/thestandardsgame. Rick Jamison serves as Executive Editor for this series and has more books in progress. Rick's blog on geek-to-geek marketing, The Listening Post, is at http://www.synopsys.com/blogs/listeningG2G.

Thank you for your purchase of this Synopsys Press book. It is available online at http://happyabout.com/synopsyspress or at other online and physical bookstores.

Phil Dworsky
Publisher, Synopsys Press
May 2010

The Ten Commandments for Effective Standards

Practical Insights for Creating Technical Standards

By Karen Bartleson
Cartoons by Rick Jamison
Foreword by Steve Mills

SYNOPSYS®
Press

Mountain View, CA, USA

Karen Bartleson
Synopsys, Inc.
Mountain View, CA
USA

Library of Congress Control Number: 2010926943
Hardcover ISBN: 978-1-61730-002-8 (1-61730-002-0)
Paperback ISBN:978-1-61730-000-4 (1-61730-000-4)
eBook ISBN: 978-1-61730-001-1 (1-61730-001-2)

Published by Synopsys, Inc., Mountain View, CA, USA
http://www.synopsys.com/synopsys_press

Publishing services provided by Happy About®
http://www.happyabout.com

Printed in the United States of America
May 2010

Trademarks

Disclaimer

All content included in *The Ten Commandments for Effective Standards* is the result of the efforts of Synopsys, Inc. Because of the possibility of human or mechanical error, neither the author, nor Synopsys, Inc., nor Happy About, nor any of their affiliates guarantees the accuracy, adequacy or completeness of any information contained herein and neither are they responsible for any errors or omissions, or for the results obtained from the use of such information. THERE ARE NO EXPRESS OR IMPLIED WARRANTIES, INCLUDING, BUT NOT LIMITED TO, WARRANTIES OF MERCHANTABILITY OR FITNESS FOR A PARTICULAR PURPOSE relating to *The Ten Commandments for Effective Standards*. In no event shall the author, Synopsys, Inc., Happy About, or their affiliates be liable for any indirect, special or consequential damages in connection with the information provided herein.

Dedication

To all my beloved friends and family, especially my husband, Brett, and children, Angie and Matthew.

Acknowledgments

Writing this book has been a unique and rewarding undertaking—one that I hope everyone with a passion or a pile of experiences to share can accomplish. Without all the people I've toiled with, agreed with, and argued with, I wouldn't have been able to conceive of a set of suggestions that could make the standardization process go a little more smoothly.

First, I'd like to thank my boss and my friend, Rich Goldman, for talking me into writing *The Ten Commandments for Effective Standards* and for his constant support of my work. I'd also like to thank Jan Anderson Willis for hiring me into Synopsys way back when as the standards program manager—a job that apparently no one wanted—and for allowing me to become one of Synopsys' first remote employees.

I want to mention a couple of colleagues whose encouragement kept me going when the task seemed too great: Ron Ploof, who got me started with blogging where I could first publicly publish my ideas on the Ten Commandments, and Rick Jamison, whose cartoons brought a special dimension to the book and had me laughing to tears at times.

Thanks to all the reviewers and endorsers for their enriching perspectives and sharp eyes, including: Dennis Brophy, Phil Dworsky, Rich Goldman, Jeff Green, Eric Huang, Jim Hughes, Yvette Huygen, Rick Jamison, Satoshi Kojima, Stan Krolikoski, Pamela Kumar, Oz Levia, David Lin, Philippe Magarshack, Gayane Markosyan, Georgia Marszalek, Shrenik Mehta, Stephen Meier, Vazgen Melikyan, Steve Mills, Nobuyuki Nishiguchi, David Peterman, Brad Pierce, Ron Ploof, Jan Rabaey, Gadi Singer, Gary Smith, Julie Stephenson, Shinji Tanabe, and Yatin Trivedi.

Of course, I must acknowledge everyone I've worked with in the standards arena, both inside and outside Synopsys. Steve Mills brought me into the IEEE Standards Association's Corporate Advisory Group and continues to mentor me in the ways of the IEEE-SA and expand my thinking about what the IEEE-SA could become. I appreciate my counterparts at other companies from whom I learned so much and who continue to keep me on my toes as we compete like crazy on products. Thanks to all the committee members like

Vassilious Gerousis, Victor Berman, and Gabe Moretti who gave me lots to think about, and the guy (whose name escapes me) who told me at a board meeting that I was either stupid or lying, and he didn't think I was stupid. Thanks also to the Accellera Board of Directors, their long-time chair, Shrenik Mehta, their administrator, Lynn Horobin, who was always there to help me when things got interesting, and Accellera's Public Relations Counsel, Georgia Marszalek. I appreciate all the votes from everyone in all kinds of elections for positions in standards organizations over the years. Plus, I'm lucky to have met @*thestandardsman* on Twitter who retweets my standards posts from my blog. And what would I have done without the little angel on my shoulder who whispered, "Don't say it!" when standards discussions took a personal turn.

I'd like to recognize the people at Happy About—Mitchell Levy, Deneene Bell, Mark Elias, Delia Colaco, and Liz Tadman—who turned Word documents into something I can hold in my hand and give as gifts at Christmastime.

Finally, I'm grateful to my family: to my mom who raised me in the Catholic Church, giving me an appreciation for the real Ten Commandments and whom I miss very much; to my kids who liked my "this is this and that is that and blah blah blah" presentations; and to my husband for putting up with my travels, workload, and stressed-out times.

My sincere apologies to anyone I've forgotten to name. You know who you are, and you know I appreciate you!

Contents

Foreword by Steve Mills

For many of us who have been around standards awhile, it is not difficult to cite case after case highlighting how standards have made incredible contributions politically, economically, and, perhaps most importantly, socially. From saving lives to improving the quality of life for everyone on the planet, standards are fundamentally good.

As with any good thing there is a dark side, a constant struggle between the forces of good and evil. The best standards are produced when the participants in the process come together in the spirit of cooperation with the objective of producing a mutually beneficial output, which optimizes the collective objectives of all involved. Unfortunately, efforts to develop standards can fall prey to the forces of evil, whose devilish antics serve to eliminate cooperation and skew the output to benefit the objectives of a limited few.

In *The Ten Commandments for Effective Standards*, Karen Bartleson takes a novel approach to showing us the way to salvation. For standards veterans, these "commandments" will serve as reminders of what is required to produce the best standards possible. For standards newcomers, they will help to outline the fundamentals of the standards process. For everyone, they are a useful resource to help keep us on the path of standards righteousness.

Steve Mills

Steve Mills[SM] (steve.mills@hp.com) has worked at Hewlett-Packard for twenty-eight years in research and development of products for the computer and telecommunications industries. He is currently Senior Architect in the Industry Standards Program Office. Prior to moving into the Standards Program Office, he managed R&D teams responsible for mid-range market and technology research, networking products for commercial servers, and the development of continuously available platforms for use in the telecommunications industry. He has actively contributed to the governance of standards development activities at the IEEE since 2001.

Mr. Mills is the IEEE-SA President-Elect for 2010, Past-Chair of the IEEE Standards Association Corporate Advisory Group, Past-Chair of the IEEE-SA Standards Board, Past-Chair of the IEEE Standards Education Committee, Chair of the IEEE-SA Standards Board Patent Committee, and member of the IEEE-SA Board of Governors. He has also served on the IEEE Educational Activities Board and the IEEE Communication Society Standards Board.

There is no god or goddess of technical standards. I hope you're not disappointed. There are, however, lessons that many of us mere mortals have learned over the years as we've participated in the creation of standards. I have been involved in real activities that were either quite successful or dismal failures. Based on my experience in these activities over the past twenty years and ideas drawn from other experts, I put together what I like to call my Ten Commandments for Effective Standards. They weren't handed down to me from on high, but instead came to me through the schools of hard knocks and good fortune. I'd like to share them with you in this book.

The Ten Commandments for Effective Standards are not laws. They have no religious connotations. They're not even standardized. They're simply observations I've made as to what works and what doesn't when technical standards are developed.

I've worked in the field of standards for quite a while. My perspectives are specific to technical standards, particularly those for integrated circuit design. I'm part of the industry known as electronic design automation (EDA), which is fundamentally computer-aided-design for chips (computer chips, not potato chips). The technical standards that we produce for the electronics industry are primarily formats, application programming interfaces, databases, methods, and electrical and mechanical interfaces.

There are countless standards in domains that are beyond my area of expertise—medicine, food, transportation, safety, and many others. In these diverse arenas, though, I strongly suspect that the principles and behaviors of standardization in my field are strikingly similar to those in other fields as well. I've spoken with people who participate in technical standards activities beyond EDA, and they seem to share the same challenges I've faced in my own industry.

In addition to working with standards, I've become a proponent of social media and online networking for business. I have included a few thoughts about its emerging presence in standards, particularly in the fourth commandment for effective standards: be truly open.

In November 2007, I started blogging about standards. My blog is called The Standards Game[1] and is hosted by my employer, Synopsys, a leading supplier of EDA products and solutions.

As I posted short versions of each of the Ten Commandments for Effective Standards on my blog, I received some insightful comments from readers. I wish to thank them, and I've incorporated many of their concepts in this book.

It's critically important that I state: I am not a lawyer nor do I play one on TV. For those of you who like initials, that's IANALNDIPOOTV. In the second commandment for effective standards—use caution when mixing patents and standards—you must know that I am not providing any legal advice. If you find yourself in a standards activity that involves intellectual property, the second commandment can give you my opinion only. The best direction to take is straight into your legal department for their advice.

The examples I've chosen to include in this book are to help illustrate the concepts covered. Most are from the EDA industry with additional examples from other technology industries that I'm familiar with. By no means am I singling out any company, organization, or individual. Instances of success, failure, good behavior, and bad behavior abound throughout the history of technical standards development.

I realized that writing a book about how to create effective standards is like writing a book on parenting. There are many ways to raise children successfully, and no two parents completely agree on a single way to do it. What's more, no two children are the same, so the same parenting techniques that worked on the first child may need to be adjusted for a second child. Producing standards is similar, so I expect you'll take exception to some of the concepts I present in this book. I welcome the conversation because there's always something more I can learn, and I find other people's viewpoints interesting and stimulating. Feel free to post comments about this book and the individual commandments on my blog in the Community section at http://www.synopsys.com/community.

I hope you enjoy *The Ten Commandments for Effective Standards.*

1 Why Standards?

Technical standards play an important role in business as well as everyday living. They provide opportunities for market growth and competition. They enable interoperability. They make consumers happier. They bring order out of chaos. Existing technical standards are being updated constantly, and new ones are being produced in increasingly greater numbers.

Where's a decent standard when you really need one?

The current definition of a technical standard from Wikipedia states: "A technical standard is an established norm or requirement. It is usually a formal document that establishes uniform engineering or technical criteria, methods, processes and practices." In other words, it's everyone doing something the same way.

Governmental bodies and international collaborative efforts recognize that standards should facilitate interoperability, support fair trade and fair competition, increase user, consumer and government confidence, and stimulate innovation.[2,3,4] These and other aspects of technical standards bring value to companies, organizations, governments, and individuals around the world.

- Interoperability means things can work together. Products from different suppliers can plug together. Data can flow from one computer application to another. Gadgets can be connected to each other without special adapters. It is the standard interfaces, agreed to and supported by different manufacturers, which make this possible.

- Standards can make markets grow. When a common interface is made available for use, competitors can develop new products around it. A standard can prevent a monopoly by giving more than one company the opportunity to create compatible products.

- Product development costs can be reduced by standards. Eliminating the work required to create an interface because a standard one is already available means a new product can be designed in less time, and in just about every business, time is money.

- Standards can fuel innovation by providing a common starting point. Shared protocols for communicating data make the Internet phenomenal. Universal audio and video formats give rise to a myriad of music, games, and movies—and products that play them.

- Consumer product purchases can be made in confidence when a standard is part of a product. Knowing that a new electronic component will connect into an existing entertainment system makes consumers more likely to buy it.

- Tasks are simpler and less error prone when standards are used. Designing a computer chip is an enormous undertaking. Writing the description of the design in standard formats instead of rewriting it in a variety of them not only saves time but also prevents errors from being introduced during rewrites.

- Finally, standards can improve communication. When everyone speaks the same language, a better understanding can result, and people can work together more easily and effectively.

2 Why "Effective" Standards?

Because of the high cost of standards development and maintenance, it's incumbent upon the participants in a standardization activity to consider more than just producing a standard for its own sake. They should be concerned about how effective the standard will be in addressing the conditions of facilitating interoperability, allowing for fair trade and fair competition, growing the market, and stimulating innovation. A technically elegant standard that doesn't serve consumer, government, and industry needs is not an effective standard.

An effective technical standard:

- solves a real problem

- is stable

- has been proven

- may already be a de facto standard

- gives a good return on investment to the industry

- provides a standards-enhanced environment

Everyday examples of effective technical standards are part of modern living. The dimensions of drill bits and automobile wheel rims are standard. Battery ratings and shapes are standard. Electrical outlets and plugs are standard (at least within a given country).

One of the best examples of an effective technical standard in the EDA industry is called the Verilog Hardware Description Language, or simply Verilog. In the early 1980s, computer chip complexity was on the rise (as it still is today). Designing chips required increasing amounts of automation, and it was becoming problematic to manually write down every single component needed in the chip's design using existing methods.

In order for a chip design engineer to represent the functionality of a computer chip so that the functionality could be automatically converted into manufacturing instructions for the chip, a new language was required in which to describe the chip's intended behavior at a higher level of abstraction than transistors or logic gates. Much in the way that a blueprint for a kitchen represents items such as a refrigerator and a stove—without showing the mechanisms inside which make them work—the higher level language for representing the chip's components would overcome the problem of describing every individual part of the chip's design.

Two inventors, Phil Moorby and Prabu Goel, developed the Verilog language.[5] The language belonged to their company, Gateway Design Automation, and was available only for their customers to use. Gateway was acquired by Cadence Design Systems in 1989, making Verilog available to Cadence's customers, too.

After nearly a decade of use in actual, successful chip designs from multiple companies, it was clear that Verilog had proven itself to be a highly effective language for describing computer chips. But a rival language had also emerged as a standard during this period of time. The hardware description language known as IEEE Standard 1076 (VHDL) was becoming popular, and it was available for everyone to use, not just Cadence's customers.

Cadence wisely decided to make Verilog an open standard language, allowing anyone to use it without restrictions. An organization known as Open Verilog International (OVI) shepherded the Verilog language through the formal standardization process and into the IEEE Standards Association, where IEEE Standard 1364 was born in 1995.[6] The IEEE Standards Association designates its standards with the year in which they are ratified or revised, so the first version of the IEEE's Verilog standard was officially called IEEE Standard 1364-1995.

From concept to de facto standard to formal standard, Verilog became one of two languages that substantially increased the productivity of chip designers. Without the ability to describe their designs at a higher level, chip designers would have been severely curtailed in producing the next generation of complex computer chips. It would have been as if they had to describe the design of an entire skyscraper, including the description of everything in the building from drinking fountains to carpet fibers.

Today, Verilog has undergone a significant step in its evolution. It is currently known as SystemVerilog, IEEE Standard 1800-2009, and it continues to enhance the chip design process by providing an even higher level at which design engineers can work.[7]

An ineffective or detrimental technical standard may have any or all of the following attributes:

- comes from formats, methods, or technology undergoing rapid transformation

- doesn't solve a widespread problem

- is created just because it's easy

- is designed only to hurt the competition

- handcuffs innovation

- gives a poor return on investment

- results in a suboptimal solution

- creates a standards-constrained environment

A standard with any one of the characteristics above can be ineffective; it doesn't require more than one to be detrimental to an industry. At best, an ineffective standard is a waste of valuable resources. At worst, it limits market growth and harms the reputation of its creators.

In the electronics industry, a pair of controversial—and ultimately ineffective—standards were created which had several of the characteristics of a detrimental standard. Known as 1076.6 IEEE Standard for VHDL Register Transfer Level (RTL) Synthesis and 1364.1-2002 IEEE Standard for Verilog Register Transfer Level Synthesis, these standards projects absorbed much time and energy, only to be finally withdrawn as unnecessary and largely unused by chip designers.[8]

One of the advanced technologies offered by EDA companies and utilized by EDA customers is called logic synthesis. This technology automatically converts a description of a chip design into a list of interconnected components that will make up the actual chip. The description of the chip is written by the design engineers in languages such as Verilog and VHDL (mentioned previously). A key requirement for EDA companies who offer logic synthesis products is to define a subset of acceptable language statements that their products can read and interpret. The subset of statements that a synthesis product can accept is called a synthesizable subset.

One EDA company, Synopsys, had been holding a large portion of the logic synthesis market for several years. Other EDA companies offered competing products, yet Synopsys maintained its market share. Technology advancements in logic synthesis came fast, as they do today.

In an attempt to fracture the logic synthesis market, an effort was started to standardize on Synopsys' synthesizable subset and some of its underlying technology for transforming chip descriptions into lists of components. Two committees, one for Verilog and one for VHDL, were created and I participated on both committees as a representative of Synopsys. Countless engineering hours were spent poring over the proposed standards. Synopsys donated its basic subsets to the committees, and the discussions continued.

Partway through the standardization process, Synopsys was granted patents on its synthesis technology. In order to avoid the risk of losing its rights to enforce these valuable patents, Synopsys immediately withdrew its participation from both of the synthesizable subset committees.[9] Knowing the risks and consequences of continued participation in these standards projects, the decision was made by Synopsys executives (and supported by its employees) to take the conservative approach which is described in the second commandment for effective standards: use caution when mixing patents and standards. It wasn't worth the chance of losing its patent rights for Synopsys to keep working on standards that weren't clearly needed by its customers.

Needless to say, the remaining committee members weren't happy. One member threatened to start an effort to invalidate the Synopsys patents. The committees proceeded, however, and eventually created two standards for synthesizable subsets. As it turned out, the industry took very little notice. What were originally thought to be needed standards were instead merely solutions without a problem. The EDA industry could have spent its resources on a more effective standardization effort.

The Ten Commandments for Effective Standards can give insights to participants in technical standards projects and provide food for thought to everyone involved in the vital field of creating standards. They summarize real activities and behaviors from standards projects and organizations which have resulted in effective standards. Here are the Ten Commandments:

1. **Cooperate on standards, compete on products.**

   This is the Golden Rule of technical standards. The essence of standardization is to provide interfaces for multiple products to work together well, while encouraging suppliers to develop the best products possible.

2. **Use caution when mixing patents and standards.**

   Perhaps the biggest challenge faced in creating technical standards is making them available for everyone to use without restrictions while protecting the intellectual property of inventors.

3. **Know when to stop.**

   Not every standards project should be completed. Not every standards project should be started. Not everyone wants to join. Timing is important, as is having the right participants.

4. **Be truly open.**

   The word "open" has many definitions. When it comes to standards, open means available to everyone, without discrimination or conditions.

5. **Realize there is no neutral party.**

   Everyone participating in a standards project has a reason for being there, whether it's obvious or not. Technical standards projects can be political.

6. **Leverage existing organizations and proven processes.**

   Reinventing the wheel isn't necessary. It's more effective to work within experienced standards-development and standards-setting organizations.

7. **Think relevance.**

   Technical standards can be expensive to produce, so it's important that they address a real need or solve a real problem.

8. **Recognize there is more than one way to create a standard.**

   Formal standards committees are just one way to create technical standards for an industry. Different methods have pros and cons.

9. **Start with contributions, not from scratch.**

   Producing standards from technology that has already been developed can speed up the standardization process and increase the quality of the resulting standard.

10. **Know that standards have technical and business aspects.**

    Getting the technical details right for a standard is necessary, and so is understanding the commercial implications.

Applying the commandments in day-to-day standards activities can help make the resulting standards more effective and easier to adopt. Perhaps these commandments can also provide a role model for industries as they leverage standards to fuel their growth.

3 The First Commandment: Cooperate on Standards, Compete on Products

If there is a Golden Rule for standardization, this is it. It is the essence of effective standards and shows maturity in industries and companies. It applies to product suppliers, of course, and there is a corollary for the organizations that produce the standards as well. It is also, arguably, the most difficult to follow—especially when the stakes (real or perceived) are high.

The need for a standard often arises after an emerging technology has been in use for a while (although there are special cases where a standard is needed in anticipation of a new product family). Different suppliers offer products that implement the technology, but in different ways. Consumers begin to experience the cost of different approaches that serve the same purpose, or they want to plug together products from different suppliers. The demand for standardization grows until the industry decides it's time to take action and produce the necessary standard(s).

When consumer demand or a critical need for a standard occurs, the first commandment for effective standards calls for suppliers to cooperate. Lack of cooperation among vendors to create a customer-demanded standard can result in two (or more) standards for which the whole industry pays a price. Product providers must either choose one standard over another—restricting their ability to sell into a broader market—or they must support two standards, which is just plain costly. The costs must be absorbed by the vendors, reducing profitability, or passed on to consumers, increasing the sales price. And there's also the annoying and inefficient situation where products simply don't work together. Consumers are then forced to buy adapters or develop their own interfaces for interoperability.

Products generally require modifications to support a standard which emerges as a result of compromise between suppliers. Some amount of engineering or R&D investment likely needs to be made in order for products to incorporate a new standard. Yet this investment should be smaller than having to support two or more separate standards.

Yes, cooperation means give-and-take for vendors. It can level the playing field and reduce proprietary lock-in when multiple vendors are able to use the same interfaces, sockets, or form-factors; but without multi-vendor backing, a standard is unlikely to survive. To mitigate the level-playing-field effect, competitors who work together on a standard can race to introduce new products in support of the standard, even developing support for the standard while it's still under development. They can also focus their limited resources on differentiating their technology rather than developing multiple interfaces. Ultimately, their customers win because they have the ability to choose the best products that suit their needs while saving money. When consumers are happy with a standard and don't feel forced to buy custom connectors or adapters, they tend to have more loyalty.

For technical standards, it can sometimes be difficult to determine during the standardization process if two or more solutions are solving the same problem. Standards developing organizations may not feel it's within their realm to judge one standard's worthiness over another, and they may allow more than one standards project to proceed. When this occurs, the marketplace is left to decide whether a single standard or multiple standards are appropriate.

A prime example of a standard that enables competition as a result of cooperation is the widely adopted USB (Universal Serial Bus). Anyone with a personal computer or electronic gadget can appreciate the simplicity of plugging devices together with common cables and ports. The USB's design is a standard that all companies can incorporate into their products. Key strengths of the USB standard are the certification required before companies are allowed to use the logo with their products, along with the marketing and breadth of support for the USB logo.

The USB standard was created in January 1996 by a core group of companies: Compaq, Digital, IBM, Intel, Northern Telecom, and Microsoft.[10] The USB specification is standardized by the Universal Serial Bus Implementers Forum (USB-IF), which has a membership of more than eight hundred companies. As stated on page three of their subscription form, the USB-IF has a code of conduct that is "designed to allow the USB-IF to comply with the law and to preserve its integrity and credibility with the public, the industry, and within the Forum."[11]

Yet the USB-IF activities have not been without disputes. Members and non-members have vied for their own solutions to be part of the standard. Companies have cried foul during the process of updating the specification. Opposing camps have formed and re-formed. Because USB is such an important standard to a colossal industry, the outcomes of any changes or enhancements to the specification have far-reaching ramifications. When standards disputes arise, the media can find that these stories make for interesting reporting.[12]

Members of the USB-IF continue, nevertheless, to collaborate and evolve the USB standard. The current incarnation of the USB standard is called Superspeed USB (USB 3.0), and it was designed to address the need for transferring data between personal computers and electronic devices at even higher speeds. Three major players—AMD, Intel, and Nvidia—had to come to an agreement as to which version of USB 3.0 to accept, allowing the Superspeed USB to become a reality.[13]

Another example of the first commandment in action is the technical standard known as the Java programming language. Java was developed at Sun Microsystems and released for public use in 1995.[14] The Java language is designed to allow software developers to write their code only once and have it run on any computer hardware. It alleviates the burden that software creators have when differences in computer hardware require that software be rewritten for each computer platform. By now, more than six million software developers have used Java to write software that runs on over four billion computers and devices.[15] The Java coffee cup logo is seen by electronic product users on a regular basis.

While Java has been a success as a technical standard, it did go through a tumultuous period in its evolution. Because of its promise as a boon to the industry and its source code being made publicly available by Sun, Java quickly became a de facto standard. In 1997, Sun Microsystems began exploring ways to turn Java into a formal standard. In 1998, Sun introduced the Java Community Process (JCP) as a method of standardization that would promote cooperation among those who wished to enhance the Java standard.[16,17]

IBM, one of Java's staunch supporters, expected Sun to deliver Java to a formal standards organization called the European Computer Manufacturers Association (Ecma International), which was Sun's desire as well. A deadline of December 1, 1999, was set to make this happen. However, the terms of the agreement between Sun and Ecma were unacceptable to Sun, so the offer was withdrawn. IBM was disappointed, believing that Java's evolution should be governed without an inordinate amount of control by Sun. IBM suggested that Java standardization proceed without Sun—a proposal that most people knew was unrealistic.[18]

In December 1999, on the day that Sun announced it had reversed its decision to give Java to Ecma, both IBM and Sun chanted the mantra that had been in the Java community since its beginning—"cooperate on standards and compete on implementation"—as the way to move forward.[19] The Java Community Process appears to have been the solution. As evidence of JCP's ongoing success, in 2003, IBM and competitor BEA Systems announced that they were collaborating to resolve technical differences between their individual Java application servers. Further, they would contribute the related specifications to JCP in support of the Java standardization process.[20]

What happens when the first commandment for effective standards is not followed? Generally, there will be two or more standards that suppliers and consumers must deal with. Time and money are wasted. Consumers may delay their purchases while they wait and see which standard emerges victorious. The marketplace is slowed until the industry comes to terms. This situation is well illustrated by the dual standards for high definition DVDs: Blu-ray and HD DVD.

As high definition TV was becoming mainstream, high definition players emerged as well. Sony developed the Blu-ray technology, and the Blu-ray Disc Association was founded by a consortium of nine electronics companies to promote the format (i.e., make it a standard). Toshiba and NEC developed their competing standard, the Advanced Optical Disc. Toshiba chaired the DVD Forum, which adopted this format and renamed it HD DVD.[21]

The "standards war" that ensued over the high definition formats lasted almost a decade. Studios and distributors were forced to choose one format over the other—rolling the dice as to which one consumers would prefer and serving only part of the market—or deciding to support both formats at double the cost to them. Some manufacturers of high definition players developed and marketed products that could play both formats. The two organizations backing their respective standards also struggled, each hoping to emerge the winner. They even tried negotiating a compromise that would bring the war to an end, but their attempts were to no avail. The marketplace was confused, and many consumers put off buying high definition players in anticipation of a single standard becoming clear.

The factors that apparently caused the market to migrate toward Blu-ray were business alliances and Sony's addition of a Blu-ray player in its PlayStation 3 video game console. As major movie studios Paramount, Universal Studios, and Warner Bros. joined Columbia Pictures, Walt Disney, and Twentieth Century Fox in supporting Blu-ray, the format pulled ahead in the race for a single standard. When DVD distributors such as Blockbuster, Netflix, and Walmart decided to move into the Blu-ray camp, the HD DVD format was doomed to dinosaur status. Sealing the fate of HD DVD was Sony's sale of more than ten million PlayStation 3 units with Blu-ray as a standard feature. In February 2008, Toshiba announced that it would no longer produce HD DVD players and recorders.[22]

The lesson from the Blu-ray/HD DVD experience is that a massive amount of resources was invested in developing two competing standards and fighting for the survival of one. Whether Toshiba, the DVD Forum, and the original supporters of HD DVD believe the investment was worth it is not clear. What is clear is that if a single format had been agreed to earlier by the Blu-ray Disc Association and the DVD Forum, the opportunity for selling high definition players and discs would have come faster. The overall industry would have seen profits sooner, and general customer satisfaction would have come more readily.

Home from the
Standards Wars

This example leads to a corollary to the first commandment that pertains to the organizations that develop the standards. The corollary is: establish a cooperative standards environment. There can be as much competition among standards-setting bodies as that between the companies that compose them. When standards bodies compete by developing similar standards in parallel, duplication of effort and wasted resources are the result. The market experiences confusion and delays.

An example of this situation occurred in the EDA industry when EDA customers—the engineers who design computer chips—needed a common format for describing how their chip designs would use less power. Lowering power consumption by computer chips means longer battery life, less environmental impact, and increased cost savings for electronic products. There were many techniques being employed by the engineers and EDA companies to represent low-power chip designs, but all of them solved essentially the same problem. Two standards emerged: the Unified Power Format from the standards-setting organization, Accellera, and the Common Power Format from another standards organization, Si2. The two organizations attempted to bring their respective standards together into a single one, delineating which organization would be responsible for the standard itself and which would drive market adoption of the sole standard. Because the two organizations had dissimilar business models and goals—and other factors such as a win-lose mentality—they couldn't reach agreement. The two standards continued, and the industry took some solace in the saying, "Two standards are better than 20."[23]

On the other hand, when standards organizations work together to determine their strengths, identify their areas of expertise, and agree to develop complementary standards, their resources are optimized and the market enjoys the benefits of a more effective standardization process.

Another instance from the EDA industry exemplifies this. A new language standard called SystemC was developed for modeling electronic systems. The organization that oversaw the creation of SystemC was named the Open SystemC Initiative (OSCI), and it managed the evolution and distribution of the language standard. The standards-setting organization mentioned above, Accellera, realized at the time that they should not compete in the standards market being served by SystemC. As a board member of Accellera, I participated in the decision making that took place. The two organizations had distinct strengths in both their business models and their standards offerings such that they could continue to coexist in a peaceful standards environment.

Cooperation on standards can promote healthy marketplaces where companies can develop their most competitive products and standards organizations can complement each other's work. This is the heart of the first commandment for effective standards.

4

The Second Commandment: Use Caution When Mixing Patents and Standards

Patents are valuable assets for companies as they prevent competitors from copying their inventions without permission. Highly successful businesses can be built upon patented ideas, so the owners of patents often go to great lengths to protect their rights to sue and receive compensation from others who implement their patented ideas without authorization.

Employee participation in standards committees can have important ramifications for a company's patent portfolio. Participants can introduce patented technology into a standard—intentionally or not. When committee procedures call for participants to announce whether they have any knowledge of patents related to the standard under development, it's important that every committee member take this seriously and not withhold information. This is the underlying reason for the second commandment for effective standards: use caution when mixing patents and standards. When a company's participants in a standardization project are aware that patents can have a profound effect, it minimizes the risks of the company facing a loss of corporate intellectual property or potentially serious charges, litigation, and court injunctions.

Simply put, it is cheating if a company's employees help develop a standard, don't reveal that the company has a patent "built into" the standard, and then assert the company's patent rights against others who use the standard. (Imagine if everyone had to pay a fee to plug a device into a USB port or an electrical outlet!) However, this does not mean that employees must be intimately familiar with their company's entire patent portfolio. Generally, standards development organizations do not require patent searches during the standardization process. But it's dangerous for employees to conceal knowledge they already have (or gain while the standard is developed) that their company's patents might have to be used in order for other companies to implement products that support the standard.

High-profile lawsuits show that companies cannot sneak patents into the standards arena without consequences. At a minimum, the standardization process is slowed down. Confusion can arise over who has rights to the patent. The standard might be abandoned while an alternative standard is pursued. Money might have to be spent on lawyers to sort things out. In the worst case, a company can lose its rights provided by its patent, rendering the patent effectively worthless.

The term often used for a patent that must be used to create products that implement a standard is essential patent. This means that in order for a product to support the standard, the patent would necessarily be infringed upon. If a company owns an essential patent and its employee(s) participates in a related standards committee, the company runs a risk of losing its right to enforce the patent. Losing this right means the company can't collect money from competitors who copy their invention or others who wish to license it, and the company can ultimately lose market share for their product. Even if a company preserves its patent rights after litigation, there can be a substantial price to be paid by the industry in terms of actual costs and lost opportunities. This is an extreme situation that does not occur often, and following the committee's rules of disclosure can ensure that it will not be an issue for employees who represent their companies in standards projects.

A patent that has to do with a company's product implementation that complies with a standard is a different animal, though. If the standard itself is free of patents, any patented product implementations that *use* the standard belong to the developer, and if the implementations are copied, the developer is entitled to assert its patent rights. Patents on the products themselves that implement a standard are actually desirable. This is good for business, and the goal of standardization is certainly not to commoditize an entire industry.

A conservative approach is simply to not mix patents with standards at all. Companies that own patents associated with a standards project can make the conscious decision to relinquish their patent rights for the good of the standard. Alternatively, they can choose to withdraw from participating in the project altogether to preserve their ability to enforce their patents (i.e., file lawsuits for infringement).

There are indeed patents and standards that coexist through special licensing schemes. When patents are involved, an effective solution that helps with broad adoption of a standard is called the patent pool. Fundamentally, a patent pool can be formed when two or more companies agree to allow each other to license their patents. Establishing a patent pool is an adjunct to the business of writing a standard. It can offer implementers a convenience in that each standards participant does not have to make separate agreements with each patent holder.

Implementing a patent pool, though, requires the participation of engineering, marketing, and management at all the companies holding the patents. Creating these arrangements is not a straightforward exercise. Legal and business issues abound, and fairness to all interested parties must be addressed. Even antitrust concerns can arise. Taking this direction should be undertaken consciously from the beginning of a standards project to minimize risk to the patent holder(s) and implementers.

There are times, however, when the investment into mixing patents and standards certainly does yield a positive return. The USB standard described in the first commandment is a good example of a successful patent pool. Adopters and promoters of the USB standard sign an agreement to license their patents to each other under RAND (Reasonable and Non-Discriminatory) terms and without royalties. Further, they agree to not assert their patent rights against others who have signed the agreement.[24] The concept for the USB patent pool is that if there is a violation of the agreement or someone tries to assert their patent rights, the USB-IF governing body will intervene and sue the offender.

A famous—or infamous—lawsuit illustrates quite well the need for the second commandment.[25,26] In 1992, Dell, the well-known global supplier of computers and services, joined a standards initiative called the Video and Electronics Standards Association (VESA). VESA is a nonprofit standards-setting organization whose membership includes most of the major suppliers of computers and software. VESA's mission is to develop and support industry-wide interface standards for the PC, workstation, and consumer electronics industries, particularly for the computer display industry. VESA operates on the premise of providing an environment for the creation, promotion, and support of open standards.

Dell participated on a VESA committee that produced a standard specification for designing a computer bus called the VESA Local Bus or VL-Bus. A computer bus is a structure that allows a computer's components to send and receive information to and from each other, including the computer's central processing unit (CPU)—its "brain"—as well as the rest of the components that make up a computer such as monitors, video boards, hard drives, and memory. The VL-Bus standard was designed so that a wide variety of components and software could connect with each other through a low-cost, high-speed bus.

Also participating on the VESA committee were other computer suppliers and Dell's competitors. Unbeknownst to VESA, Dell held a patent issued in 1991 on dual-purpose expansion slot technology, which was the same technology that was put into VESA's VL-Bus standard. The Dell participants withheld this knowledge from the VESA committee members as the standard went through its development process.

When the standard was completed, Dell claimed its exclusive rights to the patent, and Dell's lawyers cautioned the computer industry that products that implement the standard could violate the patent. If it were true and Dell was allowed to proceed, Dell would be able to sue any company in violation and potentially prevent competitors from selling products that were compliant with the standard.

After years of litigation, in 1995, the Federal Trade Commission (FTC) found that Dell had not acted in good faith and ruled that Dell could not assert their rights to the patent against the VESA standard VL-Bus.[27] The FTC decision declared that if Dell was allowed to defend its patent, this would unjustly restrict competition. Dell settled and agreed not to sue anyone over use of the standard.

Of course, Dell is not the only company in history to participate in a standardization project while attempting to preserve patent rights. The Dell case does show, however, the risks involved with mixing patents and standards.

To this day, companies work to preserve their patent rights while contributing to a standard. Complicated proposals to license and require cross-licensing are made. Policies that provide for RAND patent licensing abound. In cases of bad behavior, companies attempt to pressure or fool their competitors into relinquishing their patent rights. On the positive side, companies can properly withdraw from standards committees to preserve their patent rights. Frequently, they also make conscious decisions to forgo their rights to assert their patents in favor of a much-needed standard, which also helps their business and satisfies their customers.

Employees who participate on standards committees and represent a company that has a related patent or patent portfolio need to be aware of the significance of the situation. They should immediately ask their company's lawyer for direction. Standards organizations usually have policies to address patents, and some won't even accept donations of patented technology. The company's lawyer can interpret these policies and help the employees determine how to proceed. At one end of the spectrum, the company may decide to contribute its patents to the standards efforts. In the middle, the patents could be far enough removed from the standard as to be safe, or the company may offer to license them under RAND conditions. At the opposite end, the company might choose not to participate in the standard at all.

The ideal situation from the standpoint of the second commandment is for all standards to be free of patent issues. Either there are no essential patents to begin with, or essential patent owners are willing to offer them up or license them to others if they want to participate in standards creation. Given that the ideal is not always attainable, experts in patent law and standardization should be consulted for help.

5 The Third Commandment: Know When to Stop

It's exciting when a new standardization activity launches. There is much hope and expectation for a solution that will greatly improve productivity and lower costs for consumers and suppliers alike.

Not every standard that starts, however, is completed and adopted. Finishing the standard—through official ratification by an organization or managing entity—is important, of course. Yet it's the adoption that indicates the true measure of success for a standard. The number of consumers using the standard and the number of products that support it are the best indicators of viability and utility.

There are countless standards sitting on the shelf, unused. Resources consumed by creating these standards could certainly have been better spent. Obviously, working on a standard that isn't going anywhere is a waste. If there is a well-adopted standard that is available to everyone, it is not effective to create a competing or overlapping one.

This, the third commandment for effective standards, is: know when to stop. It also includes three corollaries: know when to start, with whom to start, and the reasons to start.

No one likes to admit defeat, but shutting down an ineffective standards activity is almost always the right decision. If an already accepted format, technology, or method that was previously unavailable to everyone is made open to all, it's important to stop working on an alternative. The EDA industry has, on several occasions, shut down standards activities that were not progressing or were superseded by a better standard. Within the IEEE Standards Association, there is a committee that oversees most of the EDA industry's IEEE standards projects. This committee is known as the DASC—Design Automation Standards Committee—and it documents within its meeting minutes all of the standards that it decides to withdraw.[28]

In one instance, Synopsys had created a de facto standard for design constraints, which are used to describe critical aspects for automating the design of a computer chip. Synopsys owned the standard and made it available for use only by its customers. The standard was a closed proprietary standard (closed proprietary standards are described further in the eighth commandment chapter). Around the year 2000, a group of Synopsys' competitors and EDA customers who wished to have interoperable products formed a standards committee to create a different standard that everyone could use. Within a short period of time, Synopsys realized the industry's demand for an open standard and made its standard available for everyone to use at no cost. Rather than continuing their standardization project, the standards committee wisely decided to stop their work and adopt the newly-opened standard.[29] Applying their efforts elsewhere was a cost-effective and appropriate course of action.

In another situation, an entire standards body decided to cease operations because its members felt they had met their goals and continuing business was not in the industry's best interest. Called the Virtual Socket Interface Alliance (VSIA), the organization formed in 1996 with a mission to dramatically enhance the productivity of the chip design community by providing leading-edge commercial and technical solutions and insight into the development, integration, and reuse of intellectual property. VSIA focused on identifying standards—either already in existence or needing to be developed—that would allow portions of chip designs to easily connect with each other. The set of standards defined or supported the socket into which the parts of the chip design would plug. After twelve years, VSIA's endeavor was wrapped up, and its unfinished standards projects were transferred to other standards development organizations, the IEEE Standards Association and the SPIRIT Consortium.[30,31] The decision to close VSIA was surely not an easy one, but it was certainly the right one.

The first corollary of the third commandment is: know when to start.

There is a proper time to start a standardization effort for a technology (or format, interface, database, methodology, etc.). Before standardizing, it should reach a certain level of maturity. Producing a standard from a technology that isn't promising is a waste of time at best and dangerous at worst.

In one case within the EDA arena, an attempt was made by the standards organization Open Verilog International to create a standard around a technology called cycle-based simulation before it was applied to real chip designs. The allure of standardizing this technology came about because a couple of standards participants thought the technology was promising. However, the standardization effort was short-lived because it became apparent that there was nothing proven—only experimental concepts—on which to base a standard. Fortunately, the standardization project was cancelled early in the process by the board of directors of Open Verilog International, of which I was a member, saving the industry from a wasted investment.

In the EDA industry, the maturity of a technology comes through usage in real design projects. Subsequently, donating the proven technology to an experienced standards development organization can be quite effective.

Overall, the quality and merit of a standard relies on it having been deployed in the "real world" with a good measure of acceptance before sending it through the standardization process.

The second corollary of the third commandment is: know with whom to start.

As described at the beginning of this chapter, in order for a standard to be successful, it must be adopted. A standards committee needs to include members who can drive markets, satisfy customer demands, and influence suppliers. It should have a fair balance of developers who implement the standard in their products and consumers who will buy the products. It also needs to have a majority of cooperative members so as not to be continually deadlocked over issues and votes.

There is an important family of standards in the computer industry—the group of networking standards known as 802 from the IEEE Standards Association. The 802 group of standards includes the widely-adopted 802.11 standard which defines wireless interfacing, popularly called WiFi. A key member of the committee that evolves 802.11 (into 802.11a, 802.11b, and so on—specifications that are part of laptop computers with a wireless feature) is Intel. A market leader and successful corporation, Intel brings its technical and business experience to the party. It's entirely possible that without Intel employees contributing to, utilizing, and promoting 802.11, the standard might never have been viable.

The third corollary of the third commandment completes the sequence: know the reasons to start.

Starting a standards effort because an individual has a pet project or because a group needs to justify its existence are *not* good reasons to warrant an investment in time, money, and brainpower.

The fundamental reason to begin a standardization effort is that a common need exists. An industry may realize that a significant cost savings will arise from a standard. A safety issue may need to be addressed in a universal manner. A small market can be expanded when multiple suppliers can build upon a single foundation. The first question that should be answered prior to the start of any standards project is, "Why are we doing this?" or, "What problem are we solving and for whom?"

6 The Fourth Commandment: Be Truly Open

When it comes to standards, the meaning of the term "open" is debated regularly. The word does have a variety of interpretations (see an entertaining list at the end of this chapter). At present, Wikipedia cites more than a dozen definitions of an open standard from different experts, organizations, and countries. Regardless of which definition is accepted, openness in a standards project is crucial to its success.

In the standards arena, "open" means available to everyone, without discrimination or restrictive conditions. Open technical standards are readily accessible and allowed to be used by anyone to develop products, offer services, and educate students. All competitors and all customers are able to utilize all or parts of the standard at their sole discretion.

"Open" includes the standardization process as well as the standards themselves. Standardization activities should be inclusive so that all stakeholders are allowed to participate in some way. When interested parties are blocked from developing, using, or modifying a standard, an open competing standard can emerge. Multiple standards that serve the same purpose place a cost burden on industry, so having the fewest number of standards that solve a common problem is the best scenario. Further, without input from all interested parties, the resulting standard may be inadequate and thereby not actually adopted.

Openness also has an aspect of being free from encumbrances such as unduly high license fees or restrictions on how the standard may be used. Recognizing this, standards organizations can have policies that require technology to be opened—that is, intellectual property rights relinquished or RAND licensing terms (often royalty-free) agreed to—before it's contributed to a standards working group. Potential participants can evaluate the technology's openness and suitability, allowing them to make informed decisions before joining the group.

Resolution 24 of the 14th Global Standards Collaboration meeting (GSC-14), held in Geneva, Switzerland, in July 2009, states that open standards include the following elements:[32]

- The standard is developed and/or approved, and maintained by a collaborative consensus-based process.

- Such process is transparent.

- Materially affected and interested parties are not excluded from such process.

- The standard is subject to RAND/FRAND Intellectual Property Right (IPR) policies which do not mandate, but may permit, at the option of the IPR holder, licensing essential intellectual property without compensation.

- The standard is published and made available to the general public under reasonable terms (including for reasonable fee or for free).

The American National Standards Institute (ANSI) defines openness as fundamentally "collaborative, balanced and consensus-based," and their open standards process has the following characteristics:[33]

- Consensus must be reached by representatives from materially affected and interested parties.

- Standards are required to undergo public reviews when any member of the public may submit comments.

- Comments from the consensus body and public review comment-
  ers must be responded to in good faith.

- An appeals process is required.

Because of their nature, open standards can be more widely adopted.
They may be richer in features when more contributions are made.
They can provide a revenue stream to the standards organization that
owns them and to the holders of any patents associated with the
standards (RAND terms must absolutely be in place, of course).

The virtues of open standardization processes include fairness, a level
playing field, and an environment conducive to consensus-building.
When everyone is allowed to contribute, the quality of the standard can
be enhanced.

A typical open environment for standardization has certain protocols.
Membership is available to all interested parties, either free or by
paying reasonable fees. Public meeting notices, anyone allowed to
contribute, undue financial barriers eliminated, rules of order imposed,
and officially published meeting minutes all make for an open standardization effort.

Open standards and processes, however, are not without risks. Too
many contributions can result in a bulky, unwieldy standard. Standards
developed by consensus can have fewer features when common
denominators are agreed upon by all participants in a standards
project, and consensus-building can take an inordinate amount of time.
It's no surprise that when rivals are working on the same standards
project, skirmishes are bound to happen. Working out differences in an
open process can take a significant amount of time and requires skilled
leadership to resolve them.

Two examples of open technical standards show the power of being truly open. The first is TCP/IP, which stands for Transmission Control Protocol/Internet Protocol. This standard is essentially a set of rules for communicating data across the Internet. The second example is HTML, the Hypertext Markup Language, which is an open format that is used to create Web pages. Without these two open standards, the Internet would not exist as we know it.

Now that Web 2.0, social media, and online networking are becoming prevalent in everyday life, the fourth commandment for effective standards—be truly open—raises a particularly interesting question in the standards world. What, if anything, should be done about bloggers, tweeters, and Facebook fans in standards committees? Should they communicate freely with people outside the committees, polling the world for ideas and comments? Or should they be made to agree to keep silent and not risk exposing the inner workings of the committee? Can a happy medium be found?

With social networks, a standards committee can extend beyond traditional boundaries to virtually anyone with an electronic device and an Internet connection. And it can do so instantaneously. Exponentially more contributions and ideas can be sent to the standards committee.

Concerns arise that standards efforts can be too open. Company secrets may be shared or incorrect (or embarrassing) statements made. Reports can be generated from within the standards committee by every participant, not just the secretary. Opinions on the direction and progress of the committee can be posted on public sites, not only the committee's official position on their Web page.

To work successfully in an open standards environment that includes participants who use social media, the answer is not censorship. Instead, members of the standards committee should use common sense and show respect for their fellow committee members. During discussions, members shouldn't say or disclose anything they don't want made public. If something is posted that is incorrect, a request for correction on the original post should be made. Alternatively, the correction can be posted on a related site. Name-calling and false accusations should not be tolerated by the committee members or their governance. These good behaviors are certainly not new. Yet they are now more essential than ever in light of the transparency that the Internet brings.

A noteworthy—and reassuring—aspect of social media usage is that it is self-correcting.[34] A blogger who continually publishes falsehoods will lose credibility rapidly and soon be ignored. A person who regularly posts inappropriate comments will be publicly dishonored and otherwise blocked from being seen. Companies are putting social media policies in place to help formalize what good behaviors should look like for their businesses, and these policies are generally appropriate for standards projects, too.

The commandment "be truly open" can be exemplified when bloggers and other social network users are part of a standardization effort. Overcoming fear of social media and online networking is the first step for the traditional standards world to take. The next step will be capitalizing on them.

Here are some definitions of "open" from Dictionary.com. Some are quite entertaining in the context of standards.

o.pen [oh'-puhn]

adjective, noun, verb

- not closed or barred at the time

- having relatively large or numerous spaces, voids, or intervals

- carried on in full view as in open warfare or open family strife

- enterable by registered voters regardless of political affiliation

- not legally repressed like open drug trafficking

- susceptible; vulnerable

- perforated or porous

- not engaged or committed

- undecided; unsettled

- unguarded by an opponent

- not yet balanced or adjusted

- articulated with a relatively large oral aperture

- not stopped by a finger

- containing neither endpoint

- so loosely woven that spaces are visible

- become receptive to knowledge, sympathy, etc.

- cut, blast, or break into

- begin a series of performances

- begin to bark, as on the scent of game

- make the first bet, bid, or lead in beginning a game

- a contest or tournament in which both amateurs and professionals may compete

Most of these definitions could fit a standards project—for better or worse!

7 The Fifth Commandment: Realize There Is No Neutral Party

For anyone involved in standardization, the fifth commandment for effective standards—realize there is no neutral party—is essential. Everyone participating in technical standards has a mission to accomplish, whether it's for business, law, safety, technology, or personal reasons. No one spends a valuable minute or a precious penny on a standards activity for which they don't care about a specific outcome.

People who are unfamiliar with standardization can be quite surprised when they come to understand how not-neutral standards participants can be. An engineer who is new to a committee can be horrified to discover that another committee member has been purposely delaying progress by throwing up smoke screens that sound like policy discussions. A product supplier can be perplexed when another supplier who has participated in developing a standard for years suddenly announces an alleged essential patent (see the second commandment—use caution when mixing patents and standards—for an explanation of an essential patent). And a customer can be shocked upon discovering shenanigans that feel like vandalism when attempting to survey the landscape about which of two standards is preferred.

An infamous example of a standards effort that became fraught with politics and how its governing organization resolved the situation is the 802.20 working group of the IEEE Standards Association.[35,36] This standard, also known as Mobile Broadband Wireless Access (MBWA), is for worldwide interoperability of mobile wireless access networks. In a simple example, it would enable a handheld smart phone user to access the Internet anywhere in the world while roaming, such as moving on a high-speed train.

Mobile devices are big business and demand for them continues to increase. Without standards that allow them to communicate with other devices and the Internet, regardless of who the service providers are or where the devices are located, the devices' usefulness would be severely limited. For example, if a person in California with a phone from AT&T couldn't send pictures to their parents in New York who have phones from Verizon, a great deal of customer dissatisfaction would arise. Thus, standards for wireless communication are crucial.

So, the stakes were high, and development of the 802.20 standard became increasingly contentious.[37] Accusations included vote-stacking, also known as block voting or dominance in which one company votes many times or uses undue influence to make other companies vote their way; artificial schedule acceleration to prevent certain members from contributing because they wouldn't have enough time to prepare their proposals; purposeful delay tactics to stall the progress of the standard; and hidden member affiliations (for instance, the chair of the working group was being paid as a consultant by one of the companies in the group). Motivating this behavior were deep-seated business interests from at least one large corporation and the desire to win a competition against a similar standard (interestingly enough, the competing standard was being developed by another IEEE working group). Members of the 802.20 working group filed appeals with the IEEE Standards Association (IEEE-SA) governance to investigate.

The IEEE-SA takes the matter of an appeal very seriously. It strives to maintain its reputation of providing a fair and open process for standards development. In the case of the 802.20 working group, the appeals and controversy were deemed by the IEEE-SA Standards Board to be more severe than was usually seen in IEEE standards working groups. The Board made the monumental decision to halt the activities of the 802.20 working group.[38] The suspension was necessary for the appeals to be resolved and a course of action to be determined that would mitigate dominance and bring transparency to the group. It was also a sort of "time out" to let tempers cool down.

After approximately three months, the IEEE-SA Standards Board approved a plan of action that allowed the 802.20 working group to proceed. The plan included removing the chair of the group and replacing him with a new chair of the Board's choice while seeking candidates to replace the other officers of the working group. The executive committee that oversees the development process of the 802 family of standards would work with the new officers to ensure that dominance would not be allowed. Members of the working group would have to disclose their affiliations and who was paying them to participate. The ballot body—the group of voters on the final draft of the standard—would be dissolved and reconstituted under the group's new leadership.[39]

The IEEE-SA Standards Board's plan for the 802.20 group was put into effect and was shown to be effective. The 802.20 standard was finally ratified by the IEEE Standards Association in June 2008—two years after its suspension was ordered. There was clearly nothing neutral about this standard's working group's activities and participants.

If an organization or an individual claims to be a "neutral third party," look deeper. This can happen in the midst of a standards dispute, and on the surface it can seem to be a means to a settlement. In actuality, the apparently neutral third party has something to gain—wasting time and effort for no reason simply does not make sense. An organization claiming neutrality may actually stand to increase membership and bring in additional funding. The self-proclaimed impartial individual may be trying to advance his reputation, positioning himself for a better job, a promotion, or an award.

In the EDA industry, there is a company called Si2 that provides technology and services that are meant to lead to standards. It is operated as a not-for-profit membership corporation, and its standards are examples of open proprietary standards that are licensed by a standards company (open proprietary standards are described further in the eighth commandment chapter). Si2 is comprised of coalitions—groups of member companies who focus on specific areas where standards may be needed. One such coalition is called Open-Access, and it offers a database interface that can be used during the process of designing computer chips. Si2 calls the OpenAccess Coalition a neutral organization of industry leaders which operates under Si2's bylaws.

However, there are several reasons that the OpenAccess Coalition isn't really a neutral organization. The industry leaders are there to help their businesses. They want to either design chips more effectively with the OpenAccess database technology, or they want to sell EDA products that support the database technology. The original technology belonged to a leading EDA company, and highly experienced engineers spend their time working to improve the standard. Valuable engineering time isn't invested without a good reason. Membership fees are required to be part of the OpenAccess Coalition. A prerequisite to joining the OpenAccess Coalition is paying membership fees to join the overarching company (Si2). Si2 has bills to pay and employees to compensate. Neutrality by Si2 would mean bankruptcy.

Of course, it's fine to be not neutral in standards projects. Every standardization effort needs participation, contribution, and dedication from those who care about producing the best possible standard. Certainly, contributors can't be asked to ignore the needs of their employers. Is it a faux pas if a member of a standards project actually points out publicly which sides he thinks members are taking or what hidden agendas he thinks are in play? Is it naïve to hope that everyone will settle down, be forthcoming, and proceed to pound out a technical standard straightaway? This is certainly an option. The person calling out these hidden agendas needs to be aware that he risks having his own strategy revealed (remember, there's no neutral party in standards). If he's willing to take the risk, the ensuing conversation can be exhilarating.

"No neutral party" does not mean everyone is polarized or biased the same way on all issues. Participants may "take sides" differently on each issue because of the way it will affect them individually and their companies. If a topic for discussion doesn't affect at least one member differently from the rest of the members of the standards project, then the subject would likely have been part of the base proposal agreed to at the outset of the project or it would have been readily accepted when the suggestion was made to include it in the standard. Addressing each topic that doesn't have 100 percent agreement among participants is the essence of a standards project.

The process by which a standards working group resolves its disagreements can be more contentious than the topics themselves. Two processes for handling differences of opinion in a group are by voting and by building consensus. When voting is used (either a simple majority, or a supermajority where more than 50 percent is required to pass), it can result in a definite polarization. Participants who didn't vote with the majority can feel alienated. Voting is an expeditious method, though, to keep the project moving along.

On the other hand, when working groups use a consensus-building process, they often experience long delays while convincing the minority objectors. Removing the last objection can be more frustrating than the debate about why such topic is beneficial to the rest of the supporters. One method to resolve conflict in a consensus-building process is to charter a task force with a small set of respected members of the working group who are less polarizing. The task force is given a chance to develop a proposal "treaty" which can then be voted on. These task force members have their reputation on the line and can often represent a collection of views. While not a full consensus, it can catalyze the working group to get closer to a unified technical standard as opposed to just voting.

The fact that there are no neutral parties in a standards project doesn't mean everyone has an ax to grind. There's no need to be paranoid and suspicious of every action taken or suggestion made in a standards committee. Being aware that participants have reasons for why they contribute to a standardization effort simply helps put things in context.

Ultimately, keeping the fifth commandment for effective standards in mind when entering into a standards activity can help bring a broader perspective and quell some of the emotions associated with any skirmishes that might erupt.

8 The Sixth Commandment: Leverage Existing Organizations and Proven Processes

Standards begin in a variety of ways for any number of reasons. Regardless of what motivates a standards effort, it's important that a process is followed to ensure a successful outcome. The overall process incorporates all the ingredients required to produce a useful standard. These can include:

- bylaws

- policies

- procedures

- guidelines

- rules of conduct

- committee and governance structure

- financial support

- administration

- Web sites

- information management tools

- press releases

- marketing activities

- market acceptance

- formal accreditation

Developing a new standardization environment is a monumental task; it is time-consuming to reinvent the wheel when starting a standards-development process. Bylaws, policies, and procedures take an inordinate amount of time if they are created from scratch. For instance, it can take a start-up standards organization six months or more just to put basic bylaws in place.

The best approach to take when launching a new standards initiative is to do so under an existing standards body—if possible—instead of creating an entirely new organization. Taking advantage of what an existing group has learned and refined over the years is a smart and efficient way to start. Existing standards bodies can offer administrative, technical, marketing, and financial resources to expedite the standards process. Trust is higher and risk is lower when standards projects work under organizations that have real standards in production use for years. In simple terms, the sixth commandment is "use best practices."

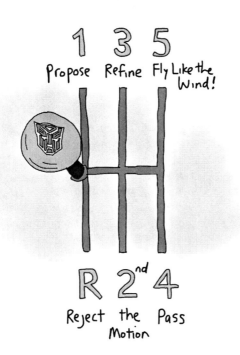

In the EDA industry, and in other industries as well, there are several standards-development and standards-setting organizations to choose from.

The most important criterion for selecting an organization to work with is finding one that is *trusted* and has a *proven track record*. Just because a standards-producing organization has a lot of money, or a large staff, or a fancy Web site, doesn't ensure that its processes will resonate with the goals of the new standards initiative. An organization that has delivered standards that are in use today by consumers and suppliers is more trustworthy than one that has been in existence for a while but hasn't delivered actual standards. In addition, it's important to select a standards organization that has a positive reputation in the industry and among consumers.

There are circumstances when a new standards body needs to be formed. For instance, if a market segment realizes that it needs standards in order to grow or stabilize, but there are no existing standards organizations that have experience with the segment's technology, it can be a good choice to start a new one. Or if a new process for developing, releasing, and maintaining a standard is devised but the available standards organizations don't have the infrastructure that will work with the process, it's appropriate to create a new standards body.

This situation can occur in fields of high tech where revolutionary inventions are the norm. In the EDA market, the desire for a standard arose in 1999 when engineers wanted to use a common language for describing their system designs.[40] They had been using the popular programming language called C++ (pronounced "see-plus-plus") for the same purpose, but many had implemented their solutions in different ways. It was inefficient for each engineering group—at the same company or different companies—to recreate descriptions or models of the parts of their systems. They wanted to be able share them with each other, and that meant a standard was needed.

A unique challenge facing this standards project was that the interested parties wanted more than just a standard language. They also wanted a standard implementation of the language—a "tool set" to be used with the language; and they wanted it to be created with the open source method of software development. As a participant in this effort to standardize a C++ solution, I found it interesting that existing standards organizations weren't set up to manage open source standards back then. Open source software (such as the Mozilla Firefox Internet browser)[41] is developed by a community of users—as opposed to a single company or individual—who use it, fix bugs, make changes, improve it, and redistribute it. The purported advantage of this method is that there are more people investing brainpower and collaborating, which makes the quality of the software higher.[42]

At the time, the existing EDA standards organizations had experience with neither open source standards nor open source implementations of the standards. The parties interested in a standard language based on C++ for designing systems decided to create their own standards body, and they borrowed best practices from companies which had successfully created open source standards. They named the standard SystemC, and they named their organization the Open SystemC Initiative (OSCI). The open source standard, SystemC, became available quickly, and product support announcements began in the summer of 2000.[43] OSCI continues to evolve the standard and its implementation. Plus, OSCI transfers stable versions of the SystemC standard to the IEEE Standards Association for its formal ratification by this accredited standards-development organization.

If, for whatever reasons, an appropriate standards organization can't be identified and a new organization is created, it's still a good idea to leverage existing organizations and proven processes. Some standards organizations are quite willing to share their bylaws and policy documents that can be used as a template, if not adopted in their entirety. Using existing principles lowers the risk of something going wrong during the standardization process and certainly shortens the time required to complete the new standard.

Another standards organization that was created in the EDA industry was called "Structure for Packaging, Integrating and Re-using IP within Tool flows," or, mercifully, just SPIRIT. When the initial group of six companies started up operations for SPIRIT, they attempted to draft a set of bylaws from the ground up.[44] It took very little time for them to realize that they could spend an inordinate amount of time on this task instead of getting to work on their standards projects, which is what they really wanted to do.

They wisely changed direction and patterned their bylaws after those from another standards organization called Accellera which had been creating EDA standards for many years. Accellera's members had refined its bylaws, policies, procedures, and other organizational rules several times as it used them over the years for real standards projects. I was a member of Accellera's Board of Directors during this time and knew that sharing the Accellera governance documents with the SPIRIT Consortium would help not only SPIRIT but also the industry at large. Because the SPIRIT Consortium took advantage of Accellera's experience, it was able to get down to the business of standardization quickly by starting with a proven set of governing documents. SPIRIT was then able to add their own unique aspects, such as four membership classes instead of two as Accellera had, to complete their bylaws.

Start-up standards organizations may find that at some point in their progression, it becomes appropriate to merge with or organize underneath an existing, mature standards body. Capitalizing on the mature body's resources such as budget and staff can be compelling enough to effect this change. This is just what happened with the SPIRIT Consortium. The organization had opened officially in July 2006. Three years later, in June 2009, SPIRIT and Accellera announced their plans to merge the two organizations.[45] The merged organization would be called Accellera, and the SPIRIT Consortium name would continue to be associated with the Consortium's ongoing standards work under the Accellera umbrella.

A final word about processes: don't violate them! Although this appears to be obvious, there are still times when participants in standardization activities fail to adhere to bylaws or stick to policies, unintentionally or not. Consequences of this behavior range from a delayed standard to banishment from participation.

Observing the sixth commandment for effective standards can save significant time and money—more of which everyone can use.

9 The Seventh Commandment: Think Relevance

In order for a technical standard to be effective, it has to fill an industry and/or end-user need. Standards are expensive and time-consuming to produce, so it's important that they give real value back to customers and suppliers. It seems obvious to say that a standards project should have a positive return on investment, yet this isn't always the case. At times, standards efforts can continue for decades with no end in sight. It may be enjoyable for the participants because the research is intriguing or the camaraderie is pleasant, but it's hard for industry to justify the cost of such an endeavor.

Pet projects without a targeted positive return on investment are not appropriate for standardization because they don't serve the market at large. It may be fun or intellectually stimulating to standardize an arcane piece of technology, but it doesn't help the community when resources are expended on efforts that aren't really necessary. Wasting time and money on minority standards that aren't widely accepted serves no purpose for the greater good.

It may be difficult to identify a minority standard from the outset. Observing whether there are a fair number of participants willing to commit their time and finances, analyzing a business plan that indicates acceptable market adoption, or assessing the amount of actual customer demand can provide guidance.

A technical standards project that is primarily an academic exercise can have some value in an educational sense. Participants can learn about standardization processes and organizations. They can explore technology that is scientifically interesting to them. Educators can use them as examples in classroom situations. Industry, however, may be reluctant to invest resources in an academic standards effort. Sending employees to meetings, paying membership fees, developing products, and marketing their goods aren't taken lightly by companies.

Unless the companies see true market relevance or a distinct competitive advantage, it's highly unlikely they will contribute to the development of a standard.

For example, in the EDA realm there is a standard whose purpose is still mostly academic since its inception in 1996. It is a computer language called Rosetta, and it can be used during the design of systems. When the development of Rosetta began, large commercial EDA companies and their customers looked on, but after a while they realized that finding a sizable market for the language was doubtful. Support from industry to standardize Rosetta waned. A handful of dedicated and interested individuals have kept the Rosetta standards project alive, nevertheless, moving it from one standards organization to another in an effort to produce a formal standard. The Rosetta standards venture persists with the hope of a few that someday it will pay off, but it's not clear if the window of opportunity will remain open.[46] Major players in the EDA industry, though, such as Cadence, Mentor, and Synopsys, have essentially ignored Rosetta as not being relevant to their businesses.[47]

Market relevance of a proposed standard can be readily recognized by clearly articulated customer demand. In situations where vendors come up with standards that customers aren't interested in and won't pay for, the standards (and the products that support them) will languish. Vendors can better spend their efforts developing valued product features instead of the standards.

On the other hand, if a standard is truly needed by customers, denial by vendors is not the right answer. At some point, if customers' demands for a standard are ignored, customers will take matters into their own hands. They may punish suppliers by removing them from preferred status. They may produce the standard independently, and then only purchase products from suppliers who support the standard. They may fund a start-up company to develop and support the standard with its products. One way or the other, customer demand for a standard almost always is satisfied.

To illustrate how customer demand drives a standards initiative—and how expensive a technical standard can be to develop—a case from the EDA industry offers some insights. Since the design of modern computer chips cannot be performed without using automation, the

EDA industry must stay on top of advances in computer chip technology. It must also solve thorny problems that arise as chips keep getting smaller and are able to perform more functions. One of the most visible challenges today in the field of chip design is devising the chips in such a way that they use less power. Consumers of electronic devices want their batteries to last longer. Computer chips become extremely hot when they are powered to run at incredibly high speeds. Sustainability and global warming are contemporary concerns. These factors and more can be addressed when chips are designed that require less power.

The ongoing hurdle of low-power chip design can be overcome through a variety of techniques developed by EDA companies and the chip designers themselves. In the 2006 time frame, the chip design community demanded a common method for describing the low-power characteristics of a chip's design. The varied, ad hoc formats that the designers were using all did essentially the same thing, and it was inefficient and risky to duplicate the low-power design information just because different EDA software products couldn't read the same description.

The chip designers—customers of the EDA companies—cried out for a standard.[48] EDA companies responded and quickly produced two competing standards in record time. One standard came from a group of EDA companies that got together and created a standard within the Accellera organization in less than six months.[49] As a member of this group, I estimated that the effort took about 7.3 man-years once the group was formed.[50] Using a representative 2006 figure of $200 per hour for fully burdened engineering costs, the group's standard cost approximately $3 million. This estimate does not include countless hours spent by the companies prior to the group's standardization work to develop their own individual formats and methods. One other EDA company created its own standard and estimated that over one hundred man-years were invested in producing it.[51] That company's standard cost more than $41 million.

Whether these figures are completely accurate or not is debatable, but they certainly show that standards development can be expensive. Despite the expense, these EDA standards were employed immediately, addressing a real market need. They are highly relevant, good examples for the seventh commandment for effective standards.

As an aside, the other part of this story is the "standards war" that erupted as a result of two standards being created. Had the first commandment for effective standards (cooperate on standards, compete on products) been observed, much of the effort that went into the battle could have been spent solving other problems for the industry, and a single, relevant standard would have been the result.

There's also a timeliness factor to consider for relevant technical standards. Standardizing an unproven or un-adopted technology is not effective. If a technology isn't seen as something everyone wants to use or it isn't well known or used enough, it's unlikely to make a relevant standard. This concept is also included in a corollary to the third commandment (know when to stop)—know when to start.

A key question to ask when starting a new standardization initiative is: "Will the standard be relevant?" If the answer is "yes," it's safe to proceed on working towards an effective standard.

Chapter 9: The Seventh Commandment: Think Relevance

10 The Eighth Commandment: Recognize There Is More than One Way to Create a Standard

When people think about how standards are made, the first image that usually enters their minds is that of a committee. Members of the committee represent various interests, and they work together to produce a detailed document that becomes the standard's specification. The specification is ultimately approved by one or more governing committees of a standards body. The standard is published and distributed by that body to be used by all interested parties. This method of producing standards—through formal standards committees—is well recognized and well used throughout the world.

There are other ways to create a standard as well. The eighth commandment for effective standards—recognize there is more than one way to create a standard—is worth considering, especially if formal standards committees can't meet all the needs of the standards producer and the community that will use the standard. The following are descriptions of standardization models, along with their advantages and disadvantages. The models may have variants, especially in regions outside the United States. There isn't a standard model for standards.

Standardization models can be grouped into four categories: formal, closed proprietary, open proprietary, and open source. The models vary by who owns the standard, how it's developed and maintained, and the availability of the standard. Each has advantages and disadvantages:

- Formal standards: developed and owned by established, reputable organizations
    - 👍 Open, fair, consensus-based, governed by policies and codes of conduct
    - 👎 Can be slow to produce, common denominator, slow to evolve
- Closed proprietary: developed and owned by a single company, available only to its customers
    - 👍 Fast to evolve, well supported, targeted to a solution
    - 👎 Discrimination, market constriction, diminished interoperability
- Open proprietary: developed and owned by a single company, available to everyone
    - 👍 Fast to evolve, well supported, targeted to a solution
    - 👎 Singly controlled evolution, potentially high fees
- Open source: developed and maintained by a community
    - 👍 Available to everyone, fast to adopt, high quality
    - 👎 Possible commercial restrictions, managed evolution, forking

Formal Standards

As mentioned above, the formal standards method is the most commonly recognized type of standardization. Formal standards, also known as de jure standards, are created, distributed, and maintained by organizations that generally are recognized as reputable, are well established, and, often, are accredited. Formal standards organizations are often simply called standards committees. Some of these organizations have obtained accreditation from higher authorities, while others have obtained trusted reputations in a given industry. Typically, they are experienced in consensus-based processes. Governed by detailed policies, procedures, and codes of conduct, the formal standards model ensures that all individuals have an equal opportunity to be heard.

The rules of operation are dictated by the committee itself or by an overseeing committee or board of directors. Establishing these rules can be an involved process because the resulting procedures must be accepted and adhered to by all members. Following the rules can be a lengthy process because all contingencies and considerations—such as fairness in the process and balanced representation—must be taken into account during the development of the standard.

For some fast-paced industries, such as the computer-chip industry, formal standards can take too long to develop because of their due processes. By the time the standard is ratified, it can be obsolete because technology advances occur while the standard is being developed. In this case, the formal standards process can be viewed as protracted and inefficient. For other industries, such as power and energy, it's perfectly acceptable for a standard to be produced after quite a few years.

Membership in a formal standards project is available to anyone who is interested in participating. Membership fees may be imposed in order to cover operating costs, but they must be reasonable so as not to preclude interested parties from joining.

In formal committees, most members sincerely want the ensuing standard to succeed. Unfortunately, individuals who want the standard to fail, often instructed to help do so by their employers, can also populate these committees, exacerbating already slow progress.

Generally, formal standards committees obtain technology donations from industry, combine the donations, and then transform them into a mutually acceptable standard. The committee members discuss the merits of the donations, obtain additional input from technical and business experts, and determine what the standard will consist of. They vote on item after item until the standard coalesces into completion.

Each participant in a formal committee is allowed one vote. Often, voting rights must be earned by attending a specified percentage of meetings or conference calls. A risk of allowing each participant to vote is that the committee can be subject to block voting. Companies have been known to send multiple individuals to standards committees in attempts to outvote each other. The governance of the standards body that oversees the committee should be vigilant and intolerant of this sort of dominance occurring in a standards project.

After the standard goes through the formal committee's process of creation and ratification, it is made available to everyone. (Nominal fees may be charged to those who wish to obtain a copy of the standard.) The standard is maintained by periodic review, revision, or reaffirmation. Formal standards organizations are well suited to managing stable, established standards that require a universal seal of approval.

Examples of formal standards are: IEEE Standards Association's 802 network family, ISO 9000, and Accellera's SystemVerilog.

A special kind of formal standard is called entity-based or corporate. This type of standardization is quite similar to the formal standardization model above with one important difference and some positive side effects. Within the standards committee where the work is performed to produce a standard, each entity is allowed just one vote. An entity can be a company, university, or an organization such as another standards body. Because each entity is permitted only a single vote, the possibility of block voting is noticeably reduced.

Higher annual fees can be paid by participants in entity-based formal standards projects. As such, entities which join the standardization effort can be more motivated to complete the project in a timely manner. Standards can even be produced in as little as a single year's time.

Additionally, when companies contribute to entity-based standards, they bring a business focus to the work that can ensure the resulting standard has market relevance. Standards are unlikely to be created from pet projects or personal interest.

As with all formal standards, entity-based standards are available to everyone when published by the formal standards organization that produces, maintains, and owns them.

Closed Proprietary Standards

Closed proprietary standards are generally viewed as not being true standards at all. Practically speaking, however, incarnations of this type of "standard" do exist and can be found readily. While the words "closed" and "proprietary" have negative connotations, a closed proprietary standard can offer tangible benefits to those entitled to use it.

A closed proprietary standard is a format, technology, or method that is developed and owned by a single company. The company invests significant resources in creating its proprietary standard and implementing it in their product(s). Applying its own technical expertise and market knowledge to create an effective standard can result in a company being able to offer a robust, speedy solution.

The owning company typically supports its closed proprietary standard very well. As the company's products evolve and it introduces new products, the standard employed by the products evolves too. The standard advances alongside the company's inventions, in a timely fashion, unencumbered by committee votes and consensus requirements. Consumers may enjoy new product choices quickly and appreciate being able to use the standard to which they're accustomed.

Closed proprietary standards are beneficial to the owner if it wishes to protect its intellectual property and trade secrets. They also protect the owner's reputation if the company's standard suffers from deficiencies that would only be worsened if it were open to interpretation and manipulation by other companies.

A closed proprietary standard is available for use only by that company's customers and selected partners. The standard is made accessible solely at the discretion of the owner. Discrimination, inequitable fees, and differing license terms are allowed because the standard belongs to the single company. It's an asset to the company and as such, the company is at liberty to capitalize on it and prevent competitors from profiting from the owner's investment.

The obvious downside to a closed proprietary standard is greatly reduced interoperability. Products from other vendors simply can't work with those from the standard's owner. This may be acceptable for a while, but eventually, consumers tend to want choices and will demand that the standard be made available to other suppliers. This situation is amplified when customers are forced to create their own interfaces or translators—at their own expense—in order to connect other products to those of the standard's owner. Over time, the possessor of a valuable, closed proprietary standard will almost always be strongly encouraged by market forces to share the standard openly.

Examples of closed proprietary standards are Apple's iTunes audio format and Cadence Design Systems' SKILL language.

Open Proprietary Standards

Open proprietary standards are also owned and controlled by a single company, yet they are accessible to anyone who wishes to use them. The advantage of this type of standard is that it protects the owner's intellectual property while promoting interoperability. Consumers realize the benefits of interoperability, suppliers can sell products into new markets, and the standard's owner can choose to profit from its asset. This standardization model ensures immediate access by everyone to well-established, well-maintained standards because the owner applies significant resources to develop, evolve, and support the standard. If the standard has true value and is widely used, it will also help the market grow.

An open proprietary standard is initially created by a single company with their own research and development investment. The owning company goes on to support, revise, and generally maintain this standard.

Evolution and enhancement of an open proprietary standard is timely and market-relevant. It is in the owner's best interest to ensure that the standard is up-to-date, useful, and high quality because the owner's customers pay for products based on the standard. The owner retains architectural control of the standard, eliminating the risk of "design by committee" and maintaining the standard's structural design principles and integrity. Another key advantage of open proprietary standards is the ease with which enhancements can be made to the standard and released to products that support it. This can be especially valuable in fast-changing technical environments.

The company owns the rights to the standard and then makes it available to the public through a licensing scheme. A license to an open proprietary standard may be expensive, inexpensive, or even free. License terms and conditions are at the discretion of the owner. However, the owner must not discriminate against licensees and must offer reasonable terms and conditions. Otherwise, the standard is actually closed proprietary, not open.

One perceived disadvantage of an open proprietary standard is that users of the standard cannot make changes to it. They can certainly send requests for modifications to the owner, but the owner decides which—if any—it wants to implement. While this may be seen as a disadvantage by competitors, it's certainly an advantage to the standard's owner who makes significant investments in the standard and is entitled to reap rewards from the expenditures.

An example of an open proprietary standard is MPEG-4, which must be licensed but is available to everyone.

Open proprietary standards can also come from companies (nonprofit or for-profit) whose sole business is producing and distributing standards. A standards company can provide a means for transferring closed proprietary standards into open proprietary standards. The standards company then owns the standards and licenses them to the public. With this model, the standards company obtains legal rights and funding from the standard's original owner and commits to promoting a standard through marketing, management, and licensing.

The standards company may institute an advisory committee to help evolve and maintain the standard. The advisory group can be made up of paid licensees, elected representatives, volunteers, and/or appointed contributors. The standards company manages and administers the advisory committee, and membership in the committee is often limited in order to better control the standard's evolution.

This type of arrangement is particularly useful for companies that do not have the resources or expertise to promote their standard to gain wide acceptance in a given market. (There are alternatives, of course. The owner of a closed proprietary standard does not have to use a standards company in order to make its standard open for widespread adoption.) The standards company can provide personnel and infrastructure for marketing campaigns surrounding the standard. It can also develop and manage the legal licensing terms and conditions.

The users of the standard may have to pay fees to the standards company, and licenses to use the standard are available to all comers. Potential buyers should be aware, however, that some standards companies may have limited resources and expertise, and may struggle to bring out a standard quickly and provide long-term support.

Licensing by a standards company brings value to an industry when the standards company understands and addresses the needs of the standard's original owner, the suppliers who want to use the standard, and the customers who consume products based on the standard.

An example of a standards company's licensed standard is Si2's Open-Access, which is used for computer chip design automation.

Open Source Standards

There is a lot of information available about open source as a means for developing and maintaining software. Applying the concepts from open source software to technical standards offers an interesting alternative to traditional standardization models. Open source standards are rapidly gaining popularity, especially in the EDA industry. In this model, a previously created standard (developed by any other means) is made available for everyone to use, modify, and enhance. It is offered to a community at large under an appropriate open source license such as Apache. The standard belongs to the community that uses it, not to its original creator.

To initiate an open source standard, the creator of the standard puts it on a Web site for anyone to download. An electronic document that describes the standard in complete detail is the minimum requirement for an open source standard. Open source standards are particularly useful, though, when they are accompanied by implementations that use the standard. Syntax checkers, kernels, and model builders are examples of implementations of technical standards.

As the open source standard is used more broadly, ideas for enhancement and corrections for bugs are discovered by the community. Since there are many people helping to find bugs and propose potential improvements, the quality of the standard can be much improved in a short time. The community that actually uses the standard sincerely wants it to succeed. Tactics that might be used to harm the standard are minimized in light of public scrutiny.

Changes to the standard are managed and redistributed by a single person, company, or entity. Proposed enhancements and corrections from the community are sent back to the managing entity for possible inclusion in a future release of the standard. The manager is responsible for ensuring the integrity of modifications and organized release of new versions of the standard. The overseer of an open source standard reserves the right to decide which community-recommended changes are introduced into the standard.

A risk exists with open source standards because community members can make their own modifications and utilize them immediately (unless the license prevents it, which is not common). The result is that the standard is now forked. In other words, two or more versions of the standard exist, and they can be incompatible with each other. Ultimately, the standard is no longer a standard and fails because everyone is doing something different.

The risk of forking is the primary reason that a single entity manages the evolution of the standard. Community members have the responsibility to send their modifications back to the managing entity so that the entire community can make use of them. Because it's in the community's interest to maintain a single standard, forking is unusual for open source standards.

An open source license is offered to potential users of the standard which explains their rights and obligations. As users agree to the license terms, downloads of the standard ensue and the standard propagates throughout a growing community of suppliers and consumers. Because the standard is readily available via the Internet, widespread adoption can occur quite quickly.

Be mindful that some open source standards restrict commercial usage and others require fees, which can be a drawback for some users. It's important that the terms of the open source license that governs the use of the standard be well understood by users of the standard. There can be legal implications, such as patent rights, that are built into the license. Generally, though, open source standards have very few restrictions and can be used free of charge.

The open source model is advantageous for existing standards, and users of the standard benefit from its fast evolution by a community of users dedicated to maintaining the quality of the standard.

Examples of open source software and standards are Linus Torvald's Linux operating system and Synopsys Design Constraints (SDC).

A variation of the open source model for standardization is known as community source. In the community source model, a group of companies oversees the evolution of the standard instead of a single managing individual or company. As with open source standards, enhancements and corrections are gathered from a community that is vested in the success of the standard. The community source model is particularly useful for emerging standards because they can evolve at Internet speed.

The community source standard is first introduced by a company or group of companies who have a need to build products around a common standard. A steering group is formed to oversee the standard's lifecycle. As with an open source standard, the community source standard is accessible to anyone, and it is used by a community of consumers and suppliers with a common incentive to improve it. Community members download the standard from the Internet, then use it, test it, and propose enhancements to it.

Proposed modifications to fix problems and improve the standard are given back to the steering group. The steering group reviews the modifications and makes decisions about their inclusion and redistribution in future versions of the standard.

Because a steering group oversees the progression of a community source standard, additional policies and procedures are required to govern the process. Voting rights, membership restrictions, and fees to cover operating costs are just a few considerations that must be addressed when the steering committee is formed. Developing the rules that govern the process takes time, and the requisite following of them throughout the standardization process also takes time, so a community source standard may be slower to emerge and evolve than an open source standard.

Examples of community source standards are Java technologies from the Java Community Process and SystemC from the Open SystemC Initiative.

Which Model to Use?

Which model is selected for a standards project depends on several factors including: maturity of the technology, rate-of-change requirements, industry and consumer demands, and business climate. The leaders of any standardization effort have the responsibility to determine the right model to bring about the best standard possible for their constituents.

11 The Ninth Commandment: Start With Contributions, Not From Scratch

Producing technical standards for widespread industry use can take time. Depending on the method used—such as those described in the eighth commandment—it can take years to complete the standardization process. Openness, fairness, and in-depth analysis are necessary, yet they are also ingredients for a lengthy course of action.

Instead of starting to create a standard with a blank sheet of paper (or computer screen in the modern era), commencing the effort with proven formats, technology, and methods greatly accelerates the process. For fast-paced industries such as consumer electronics, it's imperative that standards keep up. Computer chip technology, which is at the heart of electronic products, advances to the next generation every two years or so. A technical standard for computer chip design that takes too long to develop can be obsolete by the time it's available for use; an accelerated standardization process is mandatory for the electronics and EDA businesses. Proven technology is a valuable asset upon which to base a standard and deliver it promptly.

There's a quality aspect as well in starting a standard with technology that has been in use by consumers and suppliers. The technology has not only been given the opportunity to be shown as feasible, but it also has many, if not all, of the kinks worked out. This provides a more solid foundation upon which to develop a standard. It can improve the quality of the resulting standard because of actual testing to flush out bugs and identify missing pieces.

Proven formats and methods accelerate standards and improve their quality, but standardization is not effective unless a technology is useful for production. The owner of the proven format, technology, or method must, of course, be willing to donate it or license it to the industry for standardization. If the owner is unwilling or unable to contribute its technology towards a standard, the participants in the standardization effort may have to develop an alternative or seek out a technology donation from elsewhere. Companies that own patented technology, for instance, may wish to withhold standardizing it in favor of revenue from products that have the technology in them or revenue from licenses to the patent. Trade secrets are a competitive advantage, so it's usually not realistic for a standards working group to expect donations of recipes for secret sauces. It's common for donations of technology to be made once the technology becomes mainstream.

A technology donation can be made to a formal standards-setting organization, an ad hoc consortium, or an open source community. The owner can also provide it through a licensing scheme, but it must be made available to everyone under reasonable and non-discriminatory terms. If RAND terms are not offered by the donor, it's likely that the technology donation will be rejected either because of policy or principle.

Every mature standards organization has a policy for accepting donations of technology, formats, and methods. Companies wishing to contribute theirs to a standards effort need to carefully review, understand, and comply with the terms of the policy. If the terms are unacceptable for some reason, the standards organization may be willing to negotiate different terms that facilitate a much-desired contribution.

A single technology donation from a single company can be met with skepticism, especially within a formal standards-setting organization. Members of the organization can view this as one company attempting to standardize only their own way of doing things, to the detriment of all other competitors. When several companies make contributions to a standards effort, it makes for a rich soil in which to grow a standard that will be adopted more readily.

An interesting example of a popular standard that began with multiple technology donations is MPEG-1.[52] This is a standard for compressing video and audio without compromising their integrity. MPEG is arguably the most widely compatible audio/video format in the world today. In a collaborative effort, the Moving Picture Experts Group (MPEG) formed in 1988 to produce standard audio and video formats. Fourteen proposals for audio and fourteen for video were submitted, and the ones that proved to be most effective during testing for functionality and quality were selected to form the basis of the MPEG-1 standard. Companies and institutions donated their technologies to this successful standard, and the benefits of their contributions to the industry and consumers are immeasurable.

In the case that only one technology donation is made to a standards effort and no other contributions are forthcoming, this could indicate that there's no need for a standard. While not always the case, it could be that the only interest in creating a standard is coming from a single vendor. In this situation, other vendors will likely continue doing things their own way and the standard will not be adopted. An un-adopted standard is really not a standard at all.

In a worst-case scenario, if a technology donation is made by one company but additional technology donations from other companies are prohibited, it's a clear indicator that the single company wants to force everyone to accept only their standard. Blocking of additional technology contributions can be attempted through bending the governing rules of a standards committee, trickery, or lobbying. In most cases, this situation is not tolerated within standards organizations.

If a single company's technology is a clear leader, however, and other companies are anxious to get their hands on it, the sole technology donation can be welcomed with open arms. This is certainly true for the EDA standard language, Verilog, which is discussed in Chapter 2, "Why 'Effective' Standards?" The evolution of Verilog into the advanced version, SystemVerilog, was made possible through contributions of additional technology from several companies participating in its standardization working group.[53]

While it's entirely possible to create a technical standard from scratch, a standards working group is best served by considering whether technology donations can improve the efficiency of the standards development process and the quality of the resulting standard.

Chapter 11: The Ninth Commandment: Start With Contributions, Not From Scratch

12 The Tenth Commandment: Know That Standards Have Technical and Business Aspects

Technologists who participate in standards efforts for the first time are often quite surprised. Standardization has a large component of business interests, not just technology. Any technology at the heart of a technical standard must be sound, of course. Yet, business-related questions must be also be answered: How much modification will current products require in order to support the proposed standard? Can new products be developed cost-effectively to make use of the standard? Will it be economically feasible for consumers to switch to a new standard? Will consumers change suppliers because a standard affords them more choices? These are but a few of the business aspects that come up during the standardization process.

Business considerations can be viewed by technologists as political, confusing, and downright frightening. However, keeping in mind the fifth commandment—realize there is no neutral party—it makes sense that companies participate in standardization as a means to satisfy their customers and further their businesses. They invest their employees' time and almost always have to pay some kind of fees to join standards organizations. Customers, too, participate in standardization efforts to ensure their needs are met.

The costs paid by both consumers and suppliers can be significant, so they expect a positive return on their investment. If a standardization effort results in a technically elegant standard that doesn't help the participants' businesses, it will likely sit on the shelf, un-adopted, and the companies will find that they have wasted valuable resources. It's not unreasonable then that participants in a standardization effort need to address business concerns equally as much as technical concerns.

Standards projects that have participants working on them with both technical and business considerations are likely to be more effective than projects that focus on only one of these aspects. The International Organization for Standardization, ISO (creator of the well-known ISO 9000 family of quality standards), takes a noteworthy approach to address the need for balancing the technical and business natures of their standards. Each of their technical committees must create a business plan relating to the committee's project.[54] The plan expresses the primary goals and strategies of the committee along with the committee's structure. It includes an analysis of the current economic climate in the project's field, along with insights into social and regulatory aspects of the proposed or revised standard. Further, the plan describes any cooperation with other organizations that are part of the committee's activities. The business plan is reviewed on a regular basis by the technical committee to ensure that it remains valid as the technical standard's development progresses. The plan is also available to the public to review and submit comments.

Here are four effective ways to meet the requirement for addressing both technical and business aspects in a standards project.

1. One way to ensure that both technical and business aspects are given equal attention is for companies to find employees who have both technical expertise and business savvy. These are the people who can navigate their way through challenging technical issues within a standard under development and also through complex commercial constraints. Training and experience in both fields of technology and business can bring a valuable breadth of perception to a participant in a standards project. It's true, nevertheless, that individuals who are skilled in two clearly different areas may be hard to find.

2. A second way to address both technical and business elements of a standard is through on-the-job training. A person working on a standards project can gain expertise in the area that's not their specialty by watching and learning from others who are part of the project. While this is effective over time and builds a broader skill base for the standards participant, it's not without risk. Participants might inadvertently make mistakes while learning. Worse, they can be tripped up intentionally by competitors with more experience.

3. A third approach is for companies to consciously and actively train their employees who are destined to work on standards projects *before* they start participating in standards development. This sounds like common sense, of course, but it does require forethought and investment in a training program. Standards education isn't necessarily at the top of the list for employee development. Standards experts, no doubt, think it should be! Two organizations that perform standards development, ANSI and IEEE, have programs to improve instruction about standards for learning institutions and continuing education.[55,56]

4. A fourth way in which to achieve the right balance of technological and business considerations in a standards project is for companies to provide more than one representative to the project's working committee. Selecting participants who can bring different areas of expertise to the project can ensure a successful outcome. This does mean additional resources must be invested, but it can certainly pay off. Stating the obvious, it's important that participants from a single company coordinate their direction

outside of the working committee's meetings. It's embarrassing and disruptive to have public disagreements between people who are assumed to be in alignment. It's also important to follow any policies or procedures of the standards working group with respect to voting. For instance, two representatives from a single company may or may not each be allowed to vote.

What about individuals who participate in standards projects who are not affiliated with a company? Do they have business concerns as well as technical interests? In most cases, individuals who work on standardization efforts actually do have dual perspectives. They bring their technical proficiency and experience to help make the standard more robust and accurate. They also have business in mind, whether they realize it or not. Their "business" is themselves. They want to further their reputation, enhance their resume, work as a consultant, or foster their networks. In rare cases, there are individuals whose motivation to participate in standards is purely to contribute technically for the good of an industry or humanity. Their business, then, is fulfilling a personal mission or leaving a legacy.

13 Go Forth and Standardize

Creating standards can be challenging, interesting, and rewarding on many levels. Cooperating with competitors is fraught with trials and tribulations. Attending to both technical issues and business considerations is a real balancing act. Producing a standard that solves a real problem for an entire industry is gratifying. Imagine the satisfaction of working on a global standard such as USB, which ultimately becomes a feature on every mobile phone in the world; or visualize contributing to the Java platform that has a million Internet developers reaping its benefits.

What happens after an effective standard is produced and adopted? It is quite helpful for companies that support the standard to have frequent, well-defined interoperability activities. Regular industry gatherings can be held to test for interoperability, such as the USB-IF's "plug fests." At third-party testing facilities, such as the University of New Hampshire's InterOperability Laboratory, assessments for interoperability and compliance can be performed for a profit. Companies can put interoperability programs in place like those offered by Synopsys to exchange products with competitors for in-house compatibility checking. All of these can assure that products and services that implement the standard have the best quality and that the standard itself remains effective.

The Ten Commandments for Effective Standards can help participants in standards projects appreciate the nuances of standardization and understand how to participate skillfully in standards projects.

There's only one thing left to say. Go forth and standardize!

Glossary

802 standards. A family of standards from the IEEE-SA that are used to enable computer networks.

1076 standard. A computer language named VHDL that is used to describe an integrated circuit's functionality.

1364 standard. A computer language named Verilog that is used to describe an integrated circuit's functionality.

Accellera. A standards-setting organization for the electronic design automation (EDA) industry: http://www.accellera.org.

ad hoc consortium. A group of individuals, companies, academic institutions, and/or governmental bodies that come together to work on special area of standardization.

adoption. The actual use of a standard in products or services.

American National Standards Institute (ANSI). A nonprofit organization that oversees standards development by the private sector (non-governmental) in the United States and provides a liaison to international standards development organizations.

Apache License. A popular set of governing rules for distribution and use of open source software.

block voting. A group of voters with a common interest collaborating to affect the outcome of a vote; also known as vote stacking.

blog. A Web log; a Web site that contains a diary, journal, or running commentary.

Cadence Design Systems (Cadence). A large supplier of products used to design integrated circuits: http://www.cadence.com.

Central Processing Unit (CPU). The "brain" inside a computer that performs its primary functions.

chip design. The process of determining how a computer chip will function and how its components will be connected to each other.

closed proprietary standard. A standard that has a single owner and is not available for use without explicit permission from the owner.

Common Power Format (CPF). A format for describing how an integrated circuit can be designed to consume less power.

community source standard. An open source standard that has a group of entities overseeing its development.

computer bus. A structure that allows a computer's components to communicate with each other.

computer chip. A device inside an electronic product that makes it function.

consensus. General agreement among participants.

contribution. Ideas and documents given to a standards committee; also known as a donation (does not refer to a monetary donation).

corporate standard. A standard developed with a "one company, one vote" process and funding from participating companies.

de facto standard. A standard that is widely accepted but was not created under a formal set of rules.

de jure standard. A standard that has been ratified through a process governed by a formal set of rules.

Denali. A supplier of products used to design integrated circuits: http://www.denali.com.

design automation. Using computer software to help design integrated circuits; computer-aided design for chips.

Design Automation Standards Committee (DASC). A committee of the IEEE that sponsors and oversees most of the standards used in the field of integrated circuit design.

design constraints. A set of specific rules that helps automate the design of an integrated circuit.

dominance. Undue influence by a group of companies or individuals that suppresses a minority voice.

donation. Documented technology given to a standards organization; also known as a contribution (does not refer to a monetary contribution).

electronic design automation (EDA). The industry that provides products used by electronics engineers to create computer chip designs.

entity-based standard. A standard created with a "one entity, one vote" process; an entity can be a company, academic institution, industry consortium, or government body.

essential patent. A patent associated with a standard that must be infringed upon in order to implement a product that uses the standard.

European Computer Manufacturers Association (Ecma International). An organization that facilitates standardization for information and communications systems: http://www.ecma-international.org.

Facebook. A social networking tool originally created for college students to connect with each other, now used worldwide by people and businesses: http://www.facebook.com.

Fair, reasonable, and non-discriminatory (FRAND). See RAND.

forked standard. A standard that has been copied and modified, resulting in two (or more) different standards.

formal standard. A standard created under a set of well-defined, documented policies and procedures.

format. A specific manner in which data is represented in a computer file.

Global Standards Collaboration (GSC). An initiative that brings participants from different countries and standards organizations together to promote more effective, worldwide standards: http://www.gsc.etsi.org.

hardware description language (HDL). A computer language that is used to describe an integrated circuit's behavior and structure.

HyperText Markup Language (HTML). A computer language used to create documents and web pages for the Internet.

IEEE Standards Association (IEEE-SA). A standards-development organization of the IEEE that produces standards for the electrical and electronics industries: http://www.standards.ieee.org.

implementation. A product creation that follows a standard's specification.

Institute for Electrical and Electronic Engineers (IEEE). A professional society for engineers: http://www.ieee.org.

integrated circuit. A complex device comprised of numerous microscopic components and usually made of silicon and other materials; a computer chip.

Intel. A large supplier of integrated circuits and other electronic products: http://www.intel.com.

intellectual property. Ideas, inventions, copyrights, and trade secrets that belong to a company, individual, or institution.

intellectual property rights (IPR). Exclusive rights that protect intangible property such as ideas, inventions, music, and literature.

interface. The boundary between products where data is exchanged.

interoperability. Two or more products working together.

International Organization for Standardization (ISO). A standards-development organization that produces worldwide standards in a wide variety of areas: http://www.iso.org.

ISO 9000. A family of standards from ISO for assuring quality in management systems.

Japan Electronics and Information Technology Industries Association (JEITA). A Japanese trade association that promotes the electronics and IT industries: http://www.jeita.or.jp.

Java. A programming language originally developed by Sun Microsystems.

Java Community Process (JCP). The process by which the Java programming language is evolved and maintained.

kernel. The central or core part of a computer operating system.

Linux. A computer operating system developed and maintained with the open source model.

logic synthesis. The process of converting a general description of an integrated circuit into specific, interconnected components.

low-power chip. An integrated circuit that is specifically designed to consume less power.

McAfee. A large supplier of computer security software: http://www.mcafee.com.

Mentor Graphics. A large supplier of products used to design integrated circuits: http://www.mentor.com.

method. A process for performing a task or solving a problem.

Mitsubishi Electric. A large supplier of electronic products: http://www.mitsubishielectric.com.

model builder. A computer program that generates representations of the components that make up an integrated circuit or electronic system.

Moving Picture Experts Group (MPEG). A family of standards for compressing data for video, music, movies, etc.: http://mpeg.chiariglione.org/.

NEC Corporation. A large supplier of electronic products: http://www.nec.com.

OC New Media. A social media consulting company: http://www.ronamok.com.

OpenAccess. A database structure for storing information about an integrated circuit's behavior and configuration.

OpenAccess Coalition. The body that oversees the development and maintenance of the OpenAccess standard: http://www.si2.org/?page=69.

open proprietary standard. A standard that is owned by a single entity and is available to anyone who wants to use it.

open source standard. A standard that is developed by, distributed to, and owned by an interested community, typically under an open source agreement such as Apache.

Open SystemC Initiative (OSCI). The organization that oversees the development and promotion of the SystemC standard, prior to transferring the standard to the IEEE-SA: http://www.systemc.org.

Open Verilog International (OVI). An organization created for the purpose of marketing and evolving the Verilog language.

patent pool. An arrangement whereby a group of participants agree to license their patents to each other.

policy. An approved set of rules and/or procedures that govern the activities of a committee or an organization.

proprietary. Legally owned by an entity.

ratification. Official approval of a standard by its overseeing standardization body.

reasonable and non-discriminatory (RAND). Offering a patent license to everyone for equitable fees or for free.

register-transfer level (RTL) synthesis. The process of converting a general description of an integrated circuit into specific, interconnected components.

Semiconductor Technology Academic Research Center (STARC). A Japanese consortium that promotes the integrated circuit industry: http://www.starc.jp.

Silicon Integration Initiative (Si2). A standards company: http://www.si2.org.

social media. Internet communication channels such as Facebook, Twitter, LinkedIn, blogs, YouTube, and Flickr; also known as new media.

SPIRIT Consortium. The organization that developed standards for use in the integrated circuit design industry and subsequently merged with Accellera; abbreviation for "Structure for Packaging, Integrating and Re-using IP within Tool flows": http://www.spiritconsortium.org.

SpringSoft. A supplier of products used to design integrated circuits: http://www.springsoft.com.

standards committee. A group of people who work together to produce the specification of a standard.

standards-development organization. An organization that produces standards and is usually accredited.

standards-setting organization. An organization that produces standards and is usually not accredited.

ST Microelectronics. A large supplier of electronic products: http://www.st.com.

Synopsys. A large supplier of products used to design integrated circuits: http://www.synopsys.com.

Synopsys Design Constraints (SDC). A set of specific rules, written in a format created by Synopsys, that help automate the design of an integrated circuit.

syntax checker. A computer program that validates the accuracy of computer code.

synthesizable subset. A portion of a hardware description language that is used for logic synthesis.

SystemC. A computer language that is used to describe an electronic system or integrated circuit.

SystemVerilog. A computer language, evolved from Verilog, that is used to describe an integrated circuit's functionality.

technical standards. Documented specifications and methods for engineering a product or solution.

Texas Instruments. A large supplier of electronic products: http://www.ti.com.

Transmission Control Protocol/Internet Protocol (TCP/IP). Specific rules for sending information around computer networks and the Internet.

Unified Power Format (UPF). A format for describing how an integrated circuit can be designed to consume less power.

Universal Serial Bus (USB). A port that implements a standard for connecting electronic gadgets to each other.

USB Implementers Forum (USB-IF). The organization that promotes and supports the USB standard: http://www.usb.org.

ValleyPR. A public relations firm that represents the Accellera organization: http://www.valleypr.com.

Verilog. A computer language that is used to describe an integrated circuit's functionality.

VHDL (VHSIC—very high speed integrated circuit—hardware description language). A computer language that is used to describe an integrated circuit's functionality; an example of how acronyms can get out of control.

Video and Electronics Standards Association (VESA). An organization that creates and supports standards for the computer industry: http://www.vesa.org.

Virtual Socket Interface Alliance (VSIA). A defunct organization that identified standards for integrated circuit design: http://www.vsia.org.

Web 2.0. The second generation of the World Wide Web, characterized by multi-directional interaction between Internet users.

WiFi. Wireless communication for electronic products; short for wireless fidelity, a play on the term high fidelity or hi-fi.

Wikipedia. A free, openly-editable, Web-based encyclopedia whose entries are collaboratively written and updated by largely anonymous Internet users: http://wikipedia.org.

Notes to the Introduction

[1]Karen Bartleson, The Standards Game Blog, Synopsys, Inc., http://bit.ly/c6YuNV (http://www.synopsys.com/blogs/thestandardsgame).

Notes to Chapter 1

[2]David F. Alderman, "The U.S. Government's Role in Standards and Conformity Assessment" (presentation as a representative of Technology Services, National Institute of Standards and Technology, June 2, 2008), StrategicStandards.com, http://bit.ly/cL2idB (http://www.strategicstandards.com/files/US-GovernmentRole.pdf).

[3]GSC-14, "Resolution GSC-14/24: (IPRWG) Open Standards (reaffirmed)" (14th meeting of the GSC, Geneva, Switzerland, July 16, 2009), International Telecommunication Union, http://bit.ly/bthC1f (http://www.itu.int/ITU-T/gsc/gsc14/documents.html) (accessed April 6, 2010).

[4]Business Software Alliance, "BSA Welcomes Agreement on Compatibility of Open Standards and Intellectual Property Rights" (news release, Brussels, Belgium, September 28, 2005), BSA.org, http://bit.ly/aK8K9E (http://w3.bsa.org/eupolicy/press/newsreleases/BSA-welcomes-agreement-on-compatibility-of-Open-Standards-and-Intellectual-Property-Rights.cfm) (accessed April 6, 2010).

Notes to Chapter 2

[5]Verilog, "Verilog Resources," Verilog.com, http://www.verilog.com/ (accessed April 8, 2010).

[6]IEEE Standards Association, "IEEE Std 1364-1995 IEEE Standard Hardware Description Language Based on the Verilog® Hardware Description Language-Description," Institute of Electrical and Electronics Engineers, http://bit.ly/cO31Eq (http://standards.ieee.org/reading/ieee/std_public/description/dasc/1364-1995_desc.html) (accessed April 8, 2010).

[7]Wikipedia contributors, "SystemVerilog," *Wikipedia, The Free Encyclopedia*, http://bit.ly/dn8l9K (http://en.wikipedia.org/w/index.php?title=SystemVerilog&oldid=337061961) (accessed April 6, 2010).

[8]Institute of Electrical and Electronics Engineers, "Archived Standards: Design Automation Standards + Drafts Subscription," IEEE Xplore Digital Library, http://standards.ieee.org/catalog/olis/arch_dasc.html (accessed April 8, 2010).

[9]Judy Erkanat, "Synopsys Leaves Synthesis Group," *Electronic News*, FindArticles.com, December 16, 1996, http://bit.ly/asre0R (http://findarticles.com/p/articles/mi_m0EKF/is_n2147_v42/ai_18984127/) (accessed April 8, 2010).

Notes to Chapter 3

[10]Wikipedia contributors, "Universal Serial Bus," *Wikipedia, The Free Encyclopedia*, http://bit.ly/9eG2zr (http://en.wikipedia.org/w/index.php?title=Universal_Serial_Bus&oldid=354813416) (accessed April 8, 2010).

[11]Universal Serial Bus Implementers Forum, "Universal Serial Bus Implementers Forum Subscription Form," USB.org, http://www.usb.org/about (accessed April 8, 2010).

[12]James Delahunty, "USB-IF Sides with Apple in Pre-iTunes Dispute," *AfterDawn.com*, September 23, 2009, http://www.afterdawn.com/news/archive/19525.cfm; http://news.cnet.com/8301-13924_3-9961783-64.html (accessed April 8, 2010).

[13]Brooke Crothers, "Intel USB 3.0 Update Resolves Dispute with Nvidia, AMD," Nanotech: The Circuits Blog, *CNET.com*, August 14, 2008, http://news.cnet.com/8301-13924_3-10016929-64.html (accessed April 8, 2010).

[14]Wikipedia contributors, "Java (programming language)," *Wikipedia, The Free Encyclopedia*, http://bit.ly/a1YNgM (http://en.wikipedia.org/w/index.php?title=Java_(programming_language)&oldid=354702540) (accessed April 8, 2010).

[15]Java, "The Java History Timeline," Java.com, http://www.java.com/en/javahistory/timeline.jsp (accessed April 8, 2010).

[16]Java, "Java Community Process," JCP.org, http://jcp.org/en/home/index (accessed April 8, 2010).

[17]Wikipedia contributors, "Java Community Process," *Wikipedia, The Free Encyclopedia*, http://bit.ly/bV7ACK (http://en.wikipedia.org/w/index.php?title=Java_Community_Process&oldid=352635075) (accessed April 8, 2010).

[18]Stephen Shankland, "Sun Reverses Plan for Java Standard," *CNET.com*, December 7, 1999, http://bit.ly/d8If4V (http://news.cnet.com/Sun-reverses-plan-for-Java-standard/2100-1001_3-234061.html) (accessed April 8, 2010).

[19]Ibid.

[20]Martin LaMonica, "IBM, BEA Join on Java Strategy," *CNET.com*, November 25, 2003, http://news.cnet.com/2100-7345-5111567.html (accessed April 8, 2010).

[21]Wikipedia contributors, "Blu-ray Disc," *Wikipedia, The Free Encyclopedia*, http://bit.ly/cJCNoc (http://en.wikipedia.org/w/index.php?title=Blu-ray_Disc&oldid=354506595) (accessed April 8, 2010).

[22]Ibid.

[23]David Maliniak, "Power-Intent Standards Vie for Designers' Loyalty," *Electronic Design*, February 14, 2008, http://bit.ly/aZDKYa (http://electronicdesign.com/article/eda/power-intent-standards-vie-for-designers-loyalties.aspx) (accessed April 8, 2010).

Notes to Chapter 4

[24]Universal Serial Bus, "USB 2.0 Adopters Agreement," USB.org, http://www.usb.org/developers/docs/adopters.pdf.

[25]Sheila F. Anthony, "Antitrust and Intellectual Property Law: From Adversaries to Partners," *AIPLA Quarterly Journal*, 28, no. 1, (Winter 2000), Federal Trade Commission, http://www.ftc.gov/speeches/other/aipla.shtm (accessed April 8, 2010).

[26]David S. Bloch and Scott S. Megregian, "United States: The Antitrust Risks Associated with Manipulating the Standard-Setting Process," *Mondaq.com*, October 13, 2004, http://www.mondaq.com/unitedstates/article.asp?articleid=28999 (accessed April 8, 2010).

[27]United States Federal Trade Commission, "FTC Finds Dell Corporation Restricted Competition By Failing To Disclose Patent Rights In Standard-Setting Process," news release, June 17, 1996, StrategicStandards.com, http://www.strategicstandards.com/files/Dell.pdf.

Notes to Chapter 5

[28]Peter Ashenden, "IEEE Design Automation Standards Committee (DASC) Annual Report for 2004," March 7, 2005, http://bit.ly/cUgyxr (http://standards.computer.org/sabminutes/2005Wint/DASC-report-2004.pdf).

[29]Accellera, "Accellera Technical Committee: Version 1.3, 2Q2001," (committee summary, July 19, 2001), VHDL.org, http://www.vhdl.org/vfv/hm/att-0361/01-TC-Summary_2Q01.pdf.

[30]Virtual Socket Interface Alliance, "Legacy Documents of the VSI Alliance," VSI.org, http://www.vsi.org/.

[31]Ron Wilson, "VSIA RIP: IP Consortium Winds Down Operations," *EDN.com*, July 9, 2007, http://www.edn.com/article/CA6458282.html (accessed April 8, 2010).

Notes to Chapter 6

[32]GSC-14, "Resolution GSC-14/24: (IPRWG) Open Standards (reaffirmed)" (14th meeting of the Global Standards Collaboration, Geneva, Switzerland, July 16, 2009), International Telecommunication Union, http://www.itu.int/ITU-T/gsc/gsc14/documents.html.

[33]American National Standards Institute, "Overview of the U.S. Standardization System," ANSI.org, http://bit.ly/9NIPQP (http://www.ansi.org/about_ansi/introduction/introduction.aspx?menuid=1) (accessed April 8, 2010).

[34]David Meerman Scott, "Armed with Social Media-The U.S. Department of Defense," Web Ink Now Blog, comment posted October 13, 2009, http://www.webinknow.com/2009/10/us_dod_social_media.html (accessed April 8, 2010).

Notes to Chapter 7

[35]LR Mobile News Feed, "802.20 Suspended," *LightReading.com*, July 5, 2006, http://www.lightreading.com/document.asp?doc_id=98500 (accessed April 8, 2010).

[36]Wikipedia contributors, "IEEE 802.20," *Wikipedia, The Free Encyclopedia*, http://bit.ly/b2AzKj (http://en.wikipedia.org/w/index.php?title=IEEE_802.20&oldid=347754055) (accessed April 8, 2010).

[37]Kathy Kowalenko, "Standards Uproar Leads to Working Group Overhaul," *The Institute*, December 5, 2006, Institute of Electrical and Electronics Engineers, http://bit.ly/bBhDZu (http://www.ieee.org/portal/cms_docs/iportals/education/setf/newsitems/standardsuproar.pdf).

[38]Steve Mills, IEEE memo regarding the status of 802.20, June 15, 2006, Institute of Electrical and Electronics Engineers, http://bit.ly/9LtAYP (http://www.ieee.org/portal/cms_docs_iportals/iportals/aboutus/SAS B_802.20_Suspension_Announcement.pdf).

[39]Nancy Gohring, "WiMax Rival Gets Back on Track: IEEE to Resume Working Group for the 802.20 Broadband Wireless Technology," *InfoWorld.com*, September 20, 2006, http://bit.ly/9QlloZ (http://www.infoworld.com/t/networking/wimax-rival-gets-back-track-915) (accessed April 8, 2010).

Notes to Chapter 8

[40]SystemC, "About OSCI," SystemC.org, http://www.systemc.org/about/ (accessed April 8, 2010).

[41]Mozilla, "The Mozilla Manifesto," Mozilla.org, http://www.mozilla.org/about/manifesto.en.html (accessed April 8, 2010).

[42]Open Source Initiative, "About Open Source Initiative," OpenSource.org, http://www.opensource.org/about (accessed April 8, 2010).

[43]Richard Goering and Peter Clarke, "SystemC Proponents Weigh in with System-Level Tools," *EETimes.com*, June 5, 2000, http://bit.ly/avP7D1 (http://www.eetimes.com/conf/dac/showArticle.jhtml?articleID=17406124&kc=2443) (accessed April 8, 2010).

[44]SPIRIT Consortium, "Standard FAQs: What Is the SPIRIT Consortium and Why Was It Started?" SPIRITConsortium.org, http://www.spiritconsortium.org/about/faqs/#01 (accessed April 8, 2010).

[45]Accellera and SPIRIT Consortium, "EDA Standards Organizations Accellera and The SPIRIT Consortium Announce Plans to Merge," news release, June 11, 2009, Accellera.org, http://bit.ly/NiHf9 (http://www.accellera.org/press-room/2009/Accellera_SPIRIT_Merger061109_FINAL.pdf).

Notes to Chapter 9

[46]Gabe Moretti, "ESL: Where Are We and Where Are We Going?" *EE-Times.com*, February 9, 2009, http://bit.ly/a6ZijX (http://www.eetimes.com/news/design/showArticle.jhtml?articleID=213201621&pgno=1) (accessed April 8, 2010).

[47]Wiki contributors, "Rosetta Wiki," Information Technology and Telecommunications Center Wiki, University of Kansas, https://wiki.ittc.ku.edu/rosetta_wiki/index.php/Main_Page (accessed April 8, 2010).

[48]Richard Goering, "Accellera Launches 'Unified Power Format' Effort," *EETimes.com*, September 14, 2006, http://bit.ly/baqMe0 (http://www.eetimes.com/news/design/showArticle.jhtml?articleID=193000710) (accessed April 8, 2010).

[49]Accellera, "Accellera Announces New Unified Power Format Standard to Advance Low-Power Integrated Circuit Design," news release, Napa, California, February 28, 2007, Accellera.org, http://bit.ly/9NHN3j (http://www.accellera.org/pressroom/2007/AccelleraUPF022807_final-3.pdf).

[50]Karen Bartleson, "A Decade With Accellera," The Standards Game Blog, Synopsys, Inc., February 18, 2010, http://www.synopsysoc.org/thestandardsgame/?p=604 (accessed April 8, 2010).

[51]Dr. Chi-Ping Hsu, "Pushing Power Forward with a Common Power Format: The Process of Getting It Right," *EETimes.com*, November 5, 2006, http://bit.ly/dpSlZ9 (http://www.eetimes.com/news/design/showArticle.jhtml?articleID=193600111) (accessed April 8, 2010).

Notes to Chapter 11

[52]Wikipedia contributors, "MPEG-1," *Wikipedia, The Free Encyclopedia*, http://en.wikipedia.org/wiki/Mpeg-1 (accessed April 8, 2010).

[53]Accellera, "Accellera Applauds IEEE 1800™ SystemVerilog Standard Approval," news release, Napa, California, November 9, 2005, Accellera.org, http://bit.ly/90vpk5 (http://www.accellera.org/pressroom/2005/Accellera_Applauds_IEEE_SV_PR_110905-2_FINAL.pdf)

Notes to Chapter 12

[54]Roger Frost, "ISO Business Plans Will Reduce Waste and Increase Stakeholder Involvement in Standards' Development," International Organization for Standardization, May 27 1999, http://www.iso.org/iso/pressrelease.htm?refid=Ref759 (accessed April 8, 2010).

[55]American National Standards Institute, "Education and Training Overview," ANSI.org, http://www.ansi.org/education_trainings/overview.aspx?menuid=9 (accessed April 8, 2010).

[56]Institute of Electrical and Electronics Engineers, "Standards Education," IEEE.org, http://www.ieee.org/web/education/standards/index.html (accessed April 8, 2010).

Index

A

Accellera 26, 62, 66, 73
ad hoc consortium 83
adoption 26, 29, 35, 64, 77, 79
AMD (Advanced Micro Devices) 22
ANSI (American National Standards
  Institute) 42, 90
Apache License 77
Apple 75
AT&T 51
audio 8, 75, 83

B

BEA Systems 23
block voting 51, 73
Blockbuster 24
blog 4, 5, 6
Blu-ray 23, 24
business aspects 17, 87, 90
bylaws 53, 57, 58, 61, 62

C

Cadence Design Systems (Cadence)
  13, 65, 75
censorship 45
chip design 12, 13, 14, 25, 26, 37, 38,
  66, 77, 81
closed 37, 47, 70, 74, 75, 76, 77
Columbia Pictures 24
committee 15, 17, 27, 28, 30, 31, 32,
  33, 36, 37, 38, 39, 45, 49, 52, 55,
  57, 69, 70, 71, 72, 73, 74, 76, 77,
  80, 84, 89, 90, 91
community source 79, 80
Compaq 21

competition 7, 8, 11, 14, 21, 25, 31,
  51
computer bus 30
computer chip (chip) 4, 9, 12, 13, 25,
  37, 53, 65, 66, 77, 81
consensus 42, 43, 54, 55, 70, 74
contribution 17, 43, 45, 54, 81, 83, 84
controversy 51
cooperation 20, 21, 23, 26, 89
corollary 19, 25, 37, 38, 40, 67
corporate standard 27, 73
cost 8, 11, 20, 24, 25, 28, 31, 35, 37,
  40, 41, 63, 66, 72, 80, 87, 88
CPF (Common Power Format) 26
CPU (Central Processing Unit) 31
customer demand 38, 64, 65
cycle-based simulation 38

D

DASC (Design Automation
  Standards Committee) 36
de facto standard 12, 13, 23, 37
de jure standard 70
Dell 30, 31
design automation 4, 77
design constraints 37, 79
Digital 21
disclosure 28
dominance 51, 52, 73
donation 33, 73, 83, 84

E

Ecma International (European
  Computer Manufacturers
  Association) 23

loyalty 21

M

market share 15, 28
MBWA (Mobile Broadband Wireless Access) 50
Mentor Graphics (Mentor) 65
method 4, 8, 12, 14, 17, 23, 36, 54, 60, 61, 66, 69, 70, 74, 81, 82, 83
Microsoft 21
model builder 78
Moorby, Phil 13
Mozilla Firefox 61
MPEG (Moving Picture Experts Group) 76, 83

N

NEC (NEC Corporation) 24
Netflix 24
neutral party 16, 49, 54, 88
ninth commandment 81
Northern Telecom 21
Nvidia 22

O

open 4, 13, 16, 30, 36, 37, 41, 42, 43, 44, 45, 46, 47, 51, 60, 61, 65, 70, 74, 76, 77, 79, 83, 84
open proprietary standard 53, 75, 76
open source standard 61, 77, 78, 79, 80
OpenAccess 53, 77
OpenAccess Coalition 53
OSCI (Open SystemC Initiative) 26, 61, 80
OVI (Open Verilog International) 13, 38

P

Paramount Pictures 24
patent 5, 15, 16, 27, 28, 29, 31, 32, 33, 43, 49, 83

patent pool 29
patent rights 15, 28, 29, 31, 32, 79
phone 50, 51, 93
PlayStation 24
plug 8, 12, 20, 28, 37, 93
policies 32, 33, 42, 46, 57, 58, 62, 70, 71, 80, 91
policy 49, 61, 83
political 16, 47, 88
politics 50
processes 8, 16, 43, 54, 57, 60, 61, 62, 64, 70, 71
product 4, 8, 14, 15, 16, 19, 20, 21, 22, 24, 26, 28, 29, 31, 35, 37, 38, 41, 49, 53, 61, 64, 65, 66, 67, 74, 75, 76, 77, 79, 81, 83, 87, 93, 94
profit 25, 53, 75, 76, 93
proprietary 20, 37, 53, 70, 74, 75, 76, 77

Q

quality 17, 38, 43, 61, 70, 75, 78, 79, 82, 83, 84, 89, 94

R

RAND (reasonable and non-discriminatory terms) 29, 32, 33, 42, 43, 83
ratification 35, 61, 73
reasonable 29, 42, 43, 72, 76
risk 15, 27, 28, 29, 31, 43, 45, 54, 59, 61, 73, 76, 78, 90
ROI (return on investment) 12, 14, 63
Rosetta 65
RTL (register transfer level) 14
rules 28, 43, 44, 57, 62, 71, 80, 84

S

SDC (Synopsys Design Constraints) 79
second commandment 5, 15, 27, 30, 33, 49
seventh commandment 63, 66

About the Author

Karen Bartleson has three decades' experience in the computer chip industry. She is known for her work in the area of standards for electronic design automation and continues to serve on several committees that develop technical standards. She is the author of The Standards Game, a blog focused on the standards arena. Karen holds a BSEE from California Polytechnic State University, San Luis Obispo, California, and was the recipient of the Marie R. Pistilli Women in Design Automation Achievement Award in 2002.

About the Cartoonist

Rick Jamison is a rare entity in the corporate world. By day, he's disguised as the mild-mannered Social Media Strategist at Synopsys (as well as Executive Editor of Synopsys Press Business Series). But at sundown, he reveals his real superpower as a corporate cartoonist. Part illustrator, part subject clarifier, and part Big Business underbelly tickler, his cartoons enlighten, enliven, enrich, and entertain.

About Synopsys Press

Synopsys Press offers leading-edge educational publications written by industry experts for the business and technical communities associated with electronic product design. The Business Series offers concise, focused publications, such as *The Ten Commandments for Effective Standards* and *The Synopsys Journal*, a quarterly publication for management dedicated to covering the issues facing electronic system designers. The Technical Series publications provide immediately applicable information on technical topics for electronic system designers, with a special focus on proven industry-best practices to enable the mainstream design community to adopt leading-edge technology and methodology. The Technical Series includes the *Verification Methodology Manual for Low Power* (VMM-LP). A hallmark of both Series is the extensive peer review and input process, which leads to trusted, from-the-trenches information. Additional titles are nearing publication in both the Business and Technical series.

In addition to providing up-to-the-minute information for design professionals, Synopsys Press publications serve as textbooks for university courses, including those in the Synopsys University Program (http://www.synopsys.com/Community/UniversityProgram), which provides full undergraduate and graduate level curricula in electronic design.

For more information about Synopsys Press, to contribute feedback on any of our publications, or to submit ideas, please navigate to http://www.synopsys.com/synopsys_press.

Breinigsville, PA USA
16 May 2010
238077BV00004B/3/P